System Safety, Maintainability, and Maintenance for Engineers

The safety, maintainability, and maintenance of systems have become more important than ever before. Global competition and other factors are forcing manufacturers to produce highly safe and easily maintainable engineering systems. This means that there is a definite need for safety, maintainability, and maintenance professionals to work closely during the system design and other phases of a project, and this book will help with that.

System Safety, Maintainability, and Maintenance for Engineers presents, in a single volume, what engineers will need when designing systems from the fields of safety, maintainability, and maintenance of systems when they have to all work together on one project and it provides information that the reader will require no previous knowledge to understand. Also offered are sources in the reference section at the end of each chapter so that the reader is able to find further information if needed. For reader comprehension, examples along with their solutions are included in many chapters.

This book will be useful to many people including design engineers; system engineers; safety specialists; maintainability engineers; maintenance engineers; engineering managers; graduate and senior undergraduate students of engineering; researchers and instructors of safety, maintainability, and maintenance; and engineers-at-large.

System Safety, Maintainability, and Maintenance for Engineers

B.S. Dhillon

CRC Press
Taylor & Francis Group
Boca Raton London New York

CRC Press is an imprint of the
Taylor & Francis Group, an **informa** business

First edition published 2023
by CRC Press
6000 Broken Sound Parkway NW, Suite 300, Boca Raton, FL 33487-2742

and by CRC Press
4 Park Square, Milton Park, Abingdon, Oxon, OX14 4RN

CRC Press is an imprint of Taylor & Francis Group, LLC

ISBN: 978-1-032-42608-2 (hbk)
ISBN: 978-1-032-42988-5 (pbk)
ISBN: 978-1-003-36520-4 (ebk)

DOI: 10.1201/9781003365204

Typeset in Times
by SPi Technologies India Pvt Ltd (Straive)

Dedication

This book is affectionately dedicated to the City of Ottawa for providing excellent environment to achieve my academic goals.

Contents

Preface

Nowadays, engineering systems are an important element of world economy, and each year, billions of dollars are spent to produce and maintain various types of engineering systems around the globe. Their safety, maintainability, and maintenance have become an important concern because many of these systems are highly sophisticated and contain millions of parts. For example, a Boeing Jumbo 747 is made up of approximately 4.5 million parts including fasteners. Furthermore, various studies have indicated that for many large and sophisticated systems, maintenance and support cost accounts for as much as 60–75% of their life cycle costs and sometime much more.

Needless to say, safety, maintainability, and maintenance of such systems have become more important than ever before. Global competition and other factors are forcing manufacturers to produce highly safe and easily maintainable engineering systems.

It means that there is a definite need for safety, maintainability, and maintenance professionals to work closely during the system design and other phases. To achieve this goal effectively, it is essential that they have an understanding of each other's discipline to a certain degree. At present to the best of author's knowledge, there is no book that covers the topics of safety, maintainability, and maintenance within its framework. It means, at present, to gain knowledge of each other's specialties, these specialists must study various books, reports, or articles on each of the areas in question. This approach is time-consuming and rather difficult because of the specialized nature of the material involved.

Thus, the main objective of the book is to meet the need for a single volume that combines system safety, maintainability, and maintenance. The material covered is treated in such a manner that the reader requires no previous knowledge to understand it. The sources of most of the material presented are given in the reference section at the end of each chapter. This will be useful to readers if they desire to delve more deeply into a specific area or topic. At appropriate places, the book contains examples along with their solutions, and at the end of each chapter, there are numerous problems to test the reader's comprehension in the area.

The book is composed of 15 chapters. Chapter 1 presents various introductory aspects of safety, maintainability, and maintenance including useful sources for obtaining information on safety, maintainability, and maintenance. Chapter 2 reviews mathematical concepts considered useful to understand subsequent chapters. Some of the topics covered in the chapter are arithmetic mean, mean deviation, standard deviation, Boolean algebra laws, probability properties, probability distributions, and useful mathematical definitions. Chapter 3 presents various introductory aspects of safety including the need for safety, engineering safety goals, and engineers and safety.

Chapter 4 presents a number of methods considered useful to perform safety analysis. These methods are hazards and operability analysis, fault tree analysis, technic of operations review, failure modes and effect analysis, job safety analysis, control charts, interface safety analysis, Markov method, and safety indexes. Chapter 5 presents various important aspects of safety management and costing. Some of the topics covered in the chapter are safety management-related principles, safety department functions, safety committees, safety cost estimation models and methods, and safety cost performance measurement indexes. Chapter 6 is devoted to maintenance, software, and robot safety. Some of the topics covered in the chapter are reasons for safety-related problems in maintenance, maintenance personnel safety, software hazard causing ways, software hazard analysis methods, types of robot accidents, and robot safeguard methods.

Chapter 7 presents various introductory aspects of maintainability including maintainability versus maintenance and maintainability functions. Chapter 8 is devoted to maintainability tools and specific maintainability design-related considerations. Some of the topics covered in the chapter are cause-and-effect diagram, total quality management, maintainability design factors, simplification and accessibility, and standardization and modularization. Chapter 9 presents various important aspects of maintainability management and human factors in maintainability. Some of the topics covered in the chapter are maintainability management functions in the product life cycle, maintainability organization functions, typical human behaviors, human body measurements and sensory capabilities, and human factors-related formulas. Chapter 10 is devoted to maintainability testing and demonstration. Some of the topics covered in the chapter are planning and control requirements for maintainability testing and demonstration, maintainability test approaches, and testing methods. Chapter 11 presents various introductory aspects of engineering maintenance, including the need for maintenance, maintenance-related measures, and safety in maintenance. Chapter 12 is devoted to maintenance management and control. Some of the topics covered in the chapter are maintenance management-related principles, effective maintenance management elements, maintenance project control methods, and maintenance management control indices.

Chapter 13 presents various important aspects of preventive and corrective maintenance. Some of the topics covered in the chapter are steps for developing a preventive maintenance program, preventive maintenance measures, preventive maintenance advantages and disadvantages, corrective maintenance types, and corrective maintenance-related measures. Chapter 14 is devoted to software and robotic maintenance. Some of the topics covered in the chapter are software maintenance-related facts and figures, software maintenance types, software maintenance tools, robot maintenance requirements and types, robot parts and tools for maintenance and repair, and robot inspection. Finally, Chapter 15 presents various important aspects of reliability-centered maintenance (RCM). Some of the topics covered in the chapter are RCM goals and principles, RCM process, RCM elements, and RCM program effectiveness measurement indicators.

This book will be useful to many individuals including design engineers, system engineers, safety specialists, maintainability engineers, maintenance engineers, engineering managers, graduate and senior undergraduate students of engineering, researchers and instructors of safety, maintainability, and maintenance, and engineers-at-large.

The author is deeply indebted to many individuals including family members, colleagues, friends, and students for their inputs. The invisible contributions of my children are also appreciated. Last, but not least, I thank my wife, Rosy, my other half and friend, for typing this entire book and for timely help in proofreading.

B.S. Dhillon
University of Ottawa

Author

Dr. B.S. Dhillon is an emeritus professor in the Department of Mechanical Engineering at the University of Ottawa. He has served as a chairman/director of Mechanical Engineering Department/Engineering Management Program for over ten years at the same institution. He is the founder of the probability distribution named *Dhillon Distribution/Law/Model* by statistical researchers in their publications around the world. He has published over 377 [i.e., 224 (*70 single authored + 150 co-authored*) journals and 153 conference proceedings] articles on reliability engineering, maintainability, safety, engineering management, etc. He is or has been on the editorial boards of 14 international scientific journals. In addition, Dr. Dhillon has written 51 books (i.e., *48 single authored + 3 co-authored*) on various aspects of health care, engineering management, design, reliability, safety, and quality published by Wiley (1981), Van Nostrand (1982), Butterworth (1983), Marcel Dekker (1984), Pergamon (1986), etc. His books are being used in over 100 countries and many of them are translated into languages such as German, Russian, Chinese, Arabic, and Persian (Iranian).

He has served as the general chairman of two international conferences on reliability and quality control held in Los Angeles and Paris in 1987. Prof. Dhillon has also served as a consultant to various organizations and bodies and has many years of experience in the industrial sector. He has lectured in over 50 countries, including keynote addresses at various international scientific conferences held in North America, Europe, Asia, and Africa. In March 2004, Dr. Dhillon was a distinguished speaker at the Conf./Workshop on Surgical Errors (sponsored by White House Health and Safety Committee and Pentagon), held at the Capitol Hill (One Constitution Avenue, Washington, D.C.).

Professor Dhillon attended the University of Wales where he received a BS in electrical and electronic engineering and an MS in mechanical engineering. He received a PhD in industrial engineering from the University of Windsor.

1 Introduction

1.1 SAFETY, MAINTAINABILITY, AND MAINTENANCE HISTORY

This section presents an overview of historical developments in safety, maintainability, and maintenance, separately.

1.1.1 SAFETY

The history of the safety field may be traced back to the Code of Hammurabi (2000 BC) developed by a Babylonian ruler named Hammurabi. In modern times, in 1868, a patent was awarded for the first barrier safeguard in the United States [1]. In 1893, the Railway Safety Act was passed by the U.S. Congress, and in 1912, the cooperative Safety Congress met in Milwaukee, Illinois [1, 2].

In 1931, the first commercially available book entitled *Industrial Accident Prevention* was published [3]. In 1947, a paper entitled "Engineering for Safety" was presented to the Institute of Aeronautical Sciences [4]. It clearly emphasized the importance of designing safety in airplanes. In 1962, Exhibit 62-41 entitled "System Safety Engineering for the Development of Air Force Ballistic Missiles" was released by the U.S. Air Force (USAF). In 1970, the U.S. Congress passed the Occupational Safety and Health Act (OSHA).

Over the years, many organizations, authors, and researchers have contributed to the development of the safety field. Additional information on the history of safety is available in Ref. [5].

1.1.2 MAINTAINABILITY

An early reference to the maintainability field may be traced back to 1901 to the Army Signal Corps contract for development of the Wright brothers' airplane, in which it was clearly stated that the aircraft should be "simple to operate and maintain" [6]. In the modern context, the maintainability field's beginning may be traced back to the period between World War II and the 1950s, when various efforts directly or indirectly concerned with maintainability were initiated. One important example of these efforts is a 12-part series of articles that appeared in "Machine Design" in 1956 and covered topics such as design of electronic equipment for maintainability, recommendations for designing maintenance access in electronic equipment, and designing for installation [7].

In 1960, the USAF initiated a program for developing an effective systems approach to maintainability that ultimately resulted in the development of maintainability specification MIL-M-26512. In the latter part of the 1960s, many other military documents concerning maintainability appeared. Two examples of these documents are MIL-HDBK-472 [8] and MIL-STD-470 [9].

DOI: 10.1201/9781003365204-1

In 1960, the first commercially available book on maintainability entitled *Electronic Maintainability* appeared [10]. Additional information on the history of maintainability is available in Refs. [7, 11].

1.1.3 MAINTENANCE

Although humans have felt the need for maintaining their equipment since the beginning, the origin of modern engineering maintenance may be regarded as the development of the steam engine by James Watt (1736–1819) in 1769 in Great Britain [12]. In the United States, the magazine "Factory" first appeared in 1882 and has played a very important role in the development of the maintenance field [13]. A book on maintenance of railways was published in 1886 [14].

The term "preventive maintenance" was coined in the 1950s, and in 1957, a handbook on maintenance engineering was published [15]. Over the years, many other developments in the maintenance field have taken place, and today, many universities and other institutions offer academic programs/courses on the subject.

1.2   NEED FOR SAFETY, MAINTAINABILITY, AND MAINTENANCE

Today, safety has become a very important issue because each year, a very large number of people die and get seriously injured due to workplace-related and other accidents. For example, in the United States according to the National Safety Council (NSC), there were 93,400 deaths and a very large number of disabling injuries due to accidents in 1996 [16]. The total cost of these accidents was estimated to be approximately $121 billion. Some of the other factors that are also playing an instrumental role in demanding the need for better safety are government regulations, increasing number of lawsuits, and public pressures.

Maintainability is becoming increasingly very important because of the alarmingly high operating and support costs of systems and equipment. For example, each year, the U.S. industry spends over $300 billion on plant maintenance and operations, and for the fiscal year 1997, the operations and maintenance budget request of the U.S. Department of Defense was $79 billion [17, 18]. Thus, some of the objectives for applying maintainability engineering-related principles to systems and equipment are to reduce projected maintenance cost and time through design modifications directed at maintenance simplifications, to use maintainability data for estimating equipment availability or unavailability, and to determine labor hours and other related resources needed to perform the projected maintenance.

Since the Industrial Revolution, maintenance of engineering systems has been a continuous challenge. Although over the years impressive progress has been made in maintaining equipment in the field, maintenance of equipment is still a very challenging issue because of various factors including cost, competition, and complexity. Each year, billions of dollars are spent on engineering equipment maintenance around the globe, and it means that there is a definite need for effective asset management and maintenance practices that can positively influence success factors such as reliable delivery, quality, price, profitability, safety, and speed of innovation.

These factors clearly indicate a definite need for safety, maintainability, and maintenance professionals to work closely during the product design and operation phases. In order to achieve this goal successfully, it is absolutely essential that they have some understanding of each other's discipline. Once this goal is achieved, many of these professionals' work-associated difficulties will be reduced to a tolerable level or disappear altogether, thus resulting in more safe and maintainable or maintained systems.

1.3 TERMS AND DEFINITIONS

There are a large number of terms and definitions used in safety, maintainability, and maintenance. This section presents some of the frequently used terms and definitions in these three areas taken from the published literature [19–25]:

- **Safety:** This is conservation of human life and its effectiveness, and the prevention of damage to items as per specified mission requirements.
- **Maintainability:** The probability that a failed item will be restored to its satisfactory operational state.
- **Maintenance:** All actions necessary for retaining an item or equipment in, or restoring it to, a specified condition.
- **Safety assessment:** This is quantitative/qualitative determination of safety.
- **Safety management:** This is the accomplishment of safety through the effort of other people.
- **Accident:** This is an undesired and unplanned act.
- **Unsafe condition:** This is any condition (i.e., under the right set of conditions) that will lead to an accident.
- **Safety process:** This is a series/set of procedures followed to enable an item's safety requirements to be identified and satisfied.
- **Safety function:** This is a function carried out by items which must operate on at least a required minimum level to prevent the occurrence of accidents.
- **Overhaul:** A comprehensive inspection and restoration of a piece of equipment or an item to an acceptable level at a durability time or usage limit.
- **Logistic time:** The portion of downtime occupied by the wait for a required part or tool.
- **Serviceability:** The degree of ease or difficulty with which an item can be restored to its working condition.
- **Inspection:** The qualitative observation of an item's condition or performance.
- **Maintenance concept:** A statement of the overall concept of the product specification or policy that controls the type of maintenance action to be taken for the product under consideration.
- **Continuous task:** A task that involves some kind of tracking activity (e.g., monitoring a changing situation).
- **Active repair time:** The period of downtime when repair personnel are active to affect a repair.

- **Preventive maintenance:** All actions performed on a planned, periodic, and specific schedule to keep an item/equipment in stated working condition through the process of checking and reconditioning.
- **Corrective maintenance:** The repair or unscheduled maintenance to return items or equipment to a specified state was performed because maintenance personnel or others perceived deficiencies or failures.
- **Downtime:** The total time during which the item is not in satisfactory operating state.

1.4   USEFUL SOURCES FOR OBTAINING INFORMATION ON SAFETY, MAINTAINABILITY, AND MAINTENANCE

There are many sources for obtaining safety, maintainability, and maintenance-related information. Some of the most useful sources are presented below under a number of distinct categories.

1.4.1   JOURNALS

- Journal of Safety Research
- Professional Safety
- Accident Prevention
- Hazard Prevention
- Safety Management Journal
- Reliability Engineering and System Safety
- Air Force Safety Journal
- Safety Surveyor
- Nuclear Safety
- Accident Analysis and Prevention
- International Journal of Reliability, Quality, and Safety Engineering
- RAMS ASIA (Reliability, Availability, Maintainability, and Safety (RAMS) Quarterly Journal)
- Industrial Maintenance and Plant Operation
- Maintenance Technology
- Maintenance Journal
- Maintenance and Asset Management Journal
- PEM (Plant Engineering and Maintenance)
- Journal of Quality in Maintenance Engineering
- Journal of Software Maintenance and Evolution: Research and Practice

1.4.2   BOOKS

- Handley, W., *Industrial Safety Handbook*, McGraw-Hill, London, 1969.
- Spellman, F.R., Whiting, N.E., *Safety Engineering: Principles and Practice*, Government Institutes, Rockville, Maryland, 1999.
- Dhillon, B.S., *Engineering Safety: Fundamentals, Techniques, and Applications*, World Scientific Publishing, River Edge, New Jersey, 2003.

- Hammer, W., Price, D., *Occupational Safety Management and Engineering*, Prentice Hall, Upper Saddle River, New Jersey, 2001.
- Goetsch, D.L., *Occupational Safety and Health*, Prentice Hall, Englewood Cliffs, New Jersey, 1996.
- Goldman, A.S., Slattery, T.B., *Maintainability*, John Wiley and Sons, New York, 1964.
- Blanchard, B.S., Lowery, E.E., *Maintainability Principles and Practices*, McGraw-Hill, New York, 1969.
- Cunningham, C.E., Cox, W., *Applied Maintainability Engineering*, John Wiley and Sons, New York, 1972.
- Smith, D.J., Babb, R.H., *Maintainability Engineering*, Pitman, New York, 1973.
- Dhillon, B.S., *Engineering Maintainability*, Gulf Publishing Company, Houston, TX, 1999.
- Blanchard, B.S., Verma, D., Peterson, E.L., *Maintainability: A Key to Effective Serviceability and Maintenance Management*, John Wiley and Sons, New York, 1995.
- Niebel, B.W., *Engineering Maintenance Management*, Marcel Dekker, New York, 1994.
- Kelly, A., *Maintenance Strategy*, Butterworth-Heinemann, Oxford, UK, 1997.
- Moubray, J., *Reliability-Centered Maintenance*, Industrial Press, New York, 1997.
- Dhillon, B.S., *Engineering Maintenance: A Modern Approach*, CRC Press, Boca Raton, FL, 2002.

1.4.3 ORGANIZATIONS

- World Safety Organization, P.O. Box No. 1, Lalong Laan Building, Pasay City, Metro Manila, The Philippines.
- National Safety Council, 444 North Michigan Avenue, Chicago, Illinois, U.S.A.
- American Society of Safety Engineers, 1800 East Oakton St., Des Plaines, Illinois, U.S.A.
- National Institute for Occupational Safety and Health (NIOSH), 200 Independence Avenue, SW Washington, D.C., U.S.A.
- System Safety Society, 14252 Culver Drive, Suite A-261, Irvine, California, U.S.A.
- Society of Maintenance and Reliability Professionals, 401 N. Michigan Avenue, Chicago, IL.
- Maintenance Engineering Society of Australia (MESA), 11 National Circuit, Barton, ACT, Australia.
- Japan Institute of Plant Maintenance, Shuwa-koen3-Chome Building, 3-1-38, Shiba-Koen, Minato-Ku, Tokyo, Japan.
- Society of Logistics Engineers, 8100 Professional Place, Suite 211, Hyattsville, MD.

- The Institution of Plant Engineers, 77 Great Peter Street, London, U.K.
- American Institute of Plant Engineers, 539 South Lexington Place, Anaheim, CA.

1.4.4 DATA INFORMATION SOURCES

- System Reliability Service, Safety and Reliability Directorate, UKAEA, Wigshaw Lane, Culcheth, Warrington, U.K.
- American National Standards Institute (ANSI), 11 West 42nd Street, New York, NY 10036.
- Government Industry Data Exchange Program (GIDEP) Operations Center, Department of the Navy, Corona, CA.
- National Technical Information Service (NTIS), 5285 Port Royal Road, Springfield, VA.
- IEC 706 PT3, Guide on Maintainability of Equipment, Part III: Sections Six and Seven, Verification and Collection, Analysis and Presentation of Data, 1st ed., International Electro-Technical Commission, Geneva, Switzerland.
- RACEEMDI, Electronics Equipment Maintainability Data, Reliability Analysis Center, Rome Air Development Center, Griffis Air Force Base, Rome, NY.
- Defense Technical Information Center, DTIC-FDAC, 8725 John J. Kingman Road, Suite 0944, Fort Belvoir, VA.

PROBLEMS

1. Define the following three terms:
   - Safety
   - Maintainability
   - Maintenance
2. Discuss the history of the safety field.
3. Discuss the need for safety, maintainability, and maintenance.
4. List five most useful journals for obtaining safety-related information.
5. List at least six organizations concerned with safety and maintenance.
6. Compare maintenance with maintainability.
7. What is the difference between downtime and logistic time?
8. Define the following two terms:
   - Preventive maintenance
   - Corrective maintenance
9. Compare historical developments in safety and maintenance fields.
10. Define the following four terms:
    - Safety function
    - Accident
    - Inspection
    - Serviceability

REFERENCES

1. Goetsch, D.L., *Occupational Safety and Health*, Prentice Hall, Englewood Cliffs, NJ, 1996.
2. Hammer, W., Price, D., *Occupational Safety Management and Engineering*, Prentice Hall, Upper Saddle River, NJ, 2001.
3. Heinrich, H.W., *Industrial Accident Prevention*, 3rd ed., McGraw-Hill, New York, 1950.
4. Roland, H.E., Moriarty, B., *System Safety Engineering and Management*, John Wiley and Sons, New York, 1983.
5. Dhillon, B.S., *Engineering Safety: Fundamentals, Techniques, and Applications*, World Scientific Publishing, River Edge, NJ, 2003.
6. AMCP 706-133, *Engineering Design Handbook: Maintainability Engineering Theory and Practice*, Department of Defense, Washington, DC, 1976.
7. Retterer, B.L., Kowlaski, R.A., Maintainability: A Historical Perspective, *IEEE Transactions on Reliability*, Vol. 33, 1984, pp. 56–61.
8. MIL-HDBK-472, *Maintainability Prediction*, Department of Defense, Washington, DC, 1966.
9. MIL-STD-470, *Maintainability Program Requirements*, Department of Defense, Washington, DC, 1966.
10. Akenbrandt, F.L., Ed., *Electronic Maintainability*, Engineering Publishers, Elizabeth, NJ, 1960.
11. Dhillon, B.S., *Engineering Maintainability*, Gulf Publishing Company, Houston, TX, 1999.
12. *The Volume Library: A Modern Authoritative Reference for Home and School Use*, The South-Western Company, Nashville, TN, 1993.
13. *Factory*, McGraw-Hill, New York, 1882–1968.
14. Kirkman, M.M., *Maintenance of Railways*, C.N. Trivess Printers, Chicago, 1886.
15. Morrow, L., Ed., *Maintenance Engineering Handbook*, McGraw-Hill, New York, 1957.
16. *Accident Facts*, Report, National Safety Council, Chicago, Illinois, 1996.
17. Latino, C.J., *Hidden Treasure: Eliminating Chronic Failures Can Cut Maintenance Costs Up to 60%*, Report, Reliability Center, Hopewell, VA, 1999.
18. 1997 DOD Budget, *Potential Reductions to Operation and Maintenance Program*, U.S. General Accounting Office, Washington, DC, 1996.
19. *Dictionary of Terms Used in the Safety Profession*, 3rd ed., American Society of Safety Engineers, Des Plaines, IL, 1988.
20. MIL-STD-721, *Definitions of Effectiveness Terms for Reliability, Maintainability, Human Factors, and Safety*, Department of Defense, Washington, DC.
21. McKenna, T., Oliverson, R., *Glossary of Reliability and Maintenances Terms*, Gulf Publishing Company, Houston, TX, 1997.
22. Omdahl, T., Ed., *Reliability, Availability, and Maintainability (RAM) Dictionary*, ASQC Quality Press, Milwaukee, WI, 1988.
23. Definitions of Terms for Reliability and Maintainability, MIL-STD-721C, Department of Defense, Washington, DC, 1981.
24. Engineering Design Handbook: Maintenance Engineering Techniques, AMCP 706-132, Department of the Army, Washington, DC, 1975.
25. Policies Governing Maintenance Engineering within the Department of Defense, DOD INST.4151.12, Department of Defense, Washington, DC, 1968.

2 Safety, Maintainability, and Maintenance Mathematics

2.1 INTRODUCTION

Just like in the development of other areas of science and engineering, mathematics has also played a pivotal role in the development of safety, maintainability, and maintenance fields. The history of mathematics may be traced back to the development of our currently used number symbols, sometimes referred to as "Hindu-Arabic numeral system" in the published literature [1]. Among the early evidences of the use of these number, symbols are notches found on stone columns erected by the Scythian Emperor of India named Asoka, in around 250 BC [1].

The earliest reference to the concept of probability may be traced back to a gambler manual written by Girolamo Cardano (1501–1576) [2]. However, Pierre Fermat (1601–1665) and Blaise Pascal (1623–1662) were the first two individuals who solved independently and correctly the problem of dividing the winnings in a game of chance [1, 2]. Boolean algebra, which plays an important role in modern probability theory, is named after an English mathematician George Boole (1815–1864), who published in 1847 a pamphlet titled "The Mathematical Analysis of Logic: Being an Essay towards a Calculus of Deductive Reasoning" [1–3].

Laplace transforms, often used in the area of maintenance to find solutions to first-order differential equations, were developed by a French Mathematician named Pierre-Simon Laplace (1749–1827). Additional information on the history of mathematics is available in Refs. [1, 2]. This chapter presents basic mathematical concepts that will be useful to understand subsequent chapters of this book.

2.2 ARITHMETIC MEAN, MEAN DEVIATION, AND STANDARD DEVIATION

A set of given safety, maintainability, or maintenance data is useful only if it is analyzed effectively. More specifically, there are certain characteristics of the data that are useful for describing the nature of a given data set, thus enabling better decisions related to the data. This section presents three statistical measures considered quite useful in the area of safety, maintainability, and maintenance.

DOI: 10.1201/9781003365204-2

2.2.1 ARITHMETIC MEAN

Often, the arithmetic mean is simply referred to as mean and is defined by

$$m = \frac{\sum_{i=1}^{k} DV_i}{k} \tag{2.1}$$

where

m is the mean value.
k is the number of data values.
DV_i is the data value i, for $i = 1, 2, 3, \ldots, k$.

Example 2.1

Assume that the maintenance department of an organization inspected eight identical systems and found 5, 8, 4, 6, 2, 1, 12, and 3 defects in each system. Calculate the average number of defects per system (i.e., arithmetic mean).
By substituting the given data values into Equation (2.1), we get

$$m = \frac{5+8+4+6+2+1+12+3}{8}$$
$$= 5.125$$

Thus, the average number of defects per system is 5.125. In other words, the arithmetic mean of the given data set is 5.125.

2.2.2 MEAN DEVIATION

This is a commonly used measure of dispersion, which indicates the degree to which given data tend to spread about a mean value. Mean deviation is defined by

$$D_m = \frac{\sum_{i=1}^{k} |DV_i - m|}{k} \tag{2.2}$$

where

D_m is the mean deviation.
k is the number of data values.
DV_i is the data value i, for $i = 1, 2, 3, \ldots, k$.
m is the mean value of the given data set.
$|DV_i - m|$ is the absolute value of the deviation of DV_i from k.

Example 2.2

Calculate the mean deviation of the data provided in Example 2.1. Using the data set provided in Example 2.1 and the calculated mean value (i.e., $m = 5.125$ defects/system) in Equation (2.2), we obtain

$$
\begin{aligned}
D_m &= \frac{\begin{array}{l}|5-5.125|+|8-5.125|+|4-5.125|+|6-5.125|+|2-5.125| \\ +|1-5.125|+|12-5.125|+[3-5.125]\end{array}}{8} \\
&= \frac{0.125+2.875+1.125+0.875+3.125+4.125+6.875+2.125}{8} \\
&= 2.656
\end{aligned}
$$

Thus, the mean deviation of the Example 2.1 data set is 2.656.

2.2.3 STANDARD DEVIATION

Standard deviation is a widely used measure of dispersion of data in a given data set about the mean and is defined by

$$
\sigma = \left[\frac{\sum_{i=1}^{k}(DV_i - \mu)^2}{k} \right]^{1/2}
\tag{2.3}
$$

where

σ is the standard deviation.
μ is the mean value.
k is the number of data values.
DV_i is the data value i, for $i = 1, 2, 3, \ldots, k$.

The following three properties of the standard deviation are associated with the widely used normal distribution presented later in the chapter.

- **Property I:** 99.73% of all the data values are included between $\mu - 3\sigma$ and $\mu + 3\sigma$.
- **Property II:** 95.45% of all the data values are included between $\mu - 2\sigma$ and $\mu + 2\sigma$.
- **Property III:** 68.27% of all the data values are included between $\mu - \sigma$ and $\mu + \sigma$.

Example 2.3

Calculate the standard deviation of the data set given in Example 2.1
Using the Example 2.1 data set and the calculated mean value (i.e., $m = \mu = 5.125$) in Equation (2.3), we obtain

$$\sigma = \left[\frac{\begin{array}{l}(5-5.125)^2+(8-5.125)^2+(4-5.125)^2+(6-5.125)^2\\+(2-5.125)^2+(1-5.125)^2+(12-5.125)^2+(3-5.125)^2\end{array}}{8}\right]^{1/2}$$

$$= \left[\frac{\begin{array}{l}0.015+8.265+1.265+0.7656\\+9.765+17.015+47.265+4.515\end{array}}{8}\right]^{1/2} = 3.332$$

Thus, the standard deviation of the Example 2.1 data set is 3.332.

2.3 BOOLEAN ALGEBRA LAWS

Boolean algebra plays an important role in various types of safety, maintainability, and maintenance studies and is named after George Boole (1813–1864), a mathematician. Some of the Boolean algebra laws are as follows [3, 4]:

- **Commutative Law:**

$$A + B = B + A \tag{2.4}$$

   where
      A is an arbitrary set or event.
      B is an arbitrary set or event.
      + denotes the union of sets.

$$A.B = B.A \tag{2.5}$$

   where
      Dot (.) denotes the intersection of sets. It is to be noted that Equation (2.5) sometimes is written without the dot (e.g., AB), but it still conveys the same meaning.

- **Associative Law:**

$$(AB)C = A(BC) \tag{2.6}$$

   where
      C is an arbitrary set or event.

$$(A+B)+C = A+(B+C) \tag{2.7}$$

- **Idempotent Law:**

$$A+A = A \tag{2.8}$$

$$AA = A \tag{2.9}$$

- **Absorption Law:**

$$A(A+B) = A \tag{2.10}$$

$$A+(AB) = A \tag{2.11}$$

- **Distributive Law:**

$$(A+B)(A+C) = A+BC \tag{2.12}$$

$$A(B+C) = AB+AC \tag{2.13}$$

2.4 PROBABILITY DEFINITION AND PROPERTIES

Probability is defined as follows [5, 6]:

$$P(X) = \lim_{n \to \infty}\left[\frac{N}{n}\right] \tag{2.14}$$

where

$P(X)$ is the probability of occurrence of event X.
N is the number of times event X occurs in the n repeated experiments.

Some of the probability properties are as follows [5, 6]:

- The probability of occurrence of event, say Y, is

$$0 \le P(Y) \le 1 \tag{2.15}$$

- Probability of the sample space S is

$$P(S) = 1 \tag{2.16}$$

- Probability of the negation of the space S is

$$P(\bar{S}) = 0 \qquad\qquad (2.17)$$

where
$\bar{S}$ is the negation of the sample space S.

- The probability of occurrence and non-occurrence of an event, say Y, is always

$$P(Y) + P(\bar{Y}) = 1 \qquad\qquad (2.18)$$

where
$P(Y)$ is the probability of occurrence of event Y.
$P(\bar{Y})$ is the probability of non-occurrence of event Y.

- The probability of the union of n independent events is

$$P(Y_1 + Y_2 + ---- Y_n) = 1 - \prod_{i=1}^{n}(1 - P(Y_i)) \qquad\qquad (2.19)$$

where
$P(Y_i)$ is the probability of occurrence of event Y_i for i = 1, 2, 3, ...,n.

- The probability of the union of n mutually exclusive events is

$$P(Y_1 + Y_2 + ... + Y_n) = \sum_{i=1}^{n} P(Y_i) \qquad\qquad (2.20)$$

- The probability of an intersection of n independent events is

$$P(Y_1 Y_2 Y_3 ... Y_n) = P(Y_1) P(Y_2) P(Y_3) ... P(Y_n) \qquad\qquad (2.21)$$

2.5   MATHEMATICAL DEFINITIONS

This section presents a number of mathematical definitions considered useful for performing various types of system safety, maintainability, and maintenance studies.

2.5.1   CUMULATIVE DISTRIBUTION FUNCTION

For continuous random variables, this is expressed by [5]

$$F(t) = \int_{-\infty}^{t} f(x)\,dx \qquad\qquad (2.22)$$

where

t is time.
$f(t)$ is the probability density function.
$F(t)$ is the cumulative distribution function.

For $t = \infty$, Equation (2.22) becomes

$$F(t) = \int_{-\infty}^{\infty} f(x)\,dx = 1 \qquad (2.23)$$

It simply means that the total area under the probability density curve is equal to unity.

2.5.2 PROBABILITY DENSITY FUNCTION

This is expressed by [5, 7]

$$f(t) = \frac{dF(t)}{dt} \qquad (2.24)$$

2.5.3 EXPECTED VALUE

The expected value of a continuous random variable is expressed by

$$E(t) = m = \int_{-\infty}^{\infty} tf(t)\,dt \qquad (2.25)$$

where

$E(t)$ is the expected value of the continuous random variable t.
m is the mean value.

Similarly, the expected value, $E(t)$, of a discrete random variable t is given by

$$E(t) = \sum_{i=1}^{k} t_i f(t_i) \qquad (2.26)$$

where
k is the number of discrete values of the random variable t.

2.5.4 VARIANCE

The variance, $\sigma^2(t)$, of a random variable t is expressed by

$$\sigma^2(t) = E(t^2) - \left[E(t)\right]^2 \tag{2.27}$$

or

$$\sigma^2(t) = \int_0^\infty t^2 f(t)\, dt - m^2 \tag{2.28}$$

where

m is the mean value.

2.5.5 LAPLACE TRANSFORM

The Laplace transform of the function, $f(t)$, is expressed by

$$f(s) = \int_0^\infty f(t) e^{-st}\, dt \tag{2.29}$$

where

$f(s)$ is the Laplace transform of $f(t)$.
s is the Laplace transform variable.
t is the time variable.

Example 2.4

Obtain the Laplace transform of the function

$$f(t) = e^{-\theta t} \tag{2.30}$$

where
θ is a constant.

By inserting Equation (2.30) into Equation (2.29), we obtain

$$
\begin{aligned}
f(s) &= \int_0^\infty e^{-\theta t} e^{-st}\, dt \\
&= \int_0^\infty e^{-(s+\theta)t}\, dt \\
&= \frac{1}{(s+\theta)}
\end{aligned}
\tag{2.31}
$$

TABLE 2.1

Laplace Transforms of Some Functions

No.	$f(t)$	$f(s)$
1	k (a constant)	$\dfrac{k}{s}$
2	t^n, for $n = 0, 1, 2, 3...$	$\dfrac{n!}{s^{n+1}}$
3	t	$\dfrac{1}{s^2}$
4	$\dfrac{df(t)}{dt}$	$sf(s)-f(0)$
5	$te^{-\lambda t}$	$\dfrac{1}{(s+\lambda)^2}$
6	$\theta_1 f_1(t) + \theta_2 f_2(t)$	$\theta_1 f_1(s) + \theta_2 f_2(s)$
7	$e^{-\lambda t}$	$\dfrac{1}{(s+\lambda)}$

2.5.6 LAPLACE TRANSFORMS OF COMMON FUNCTIONS

Laplace transforms of some commonly occurring functions in system safety, maintainability, and maintenance analysis-related studies are presented in Table 2.1 [8, 9].

2.5.7 LAPLACE TRANSFORM: FINAL-VALUE THEOREM

If the following limits exist, then, the final-value theorem may be defined as

$$\lim_{t \to \infty} f(t) = \lim_{s \to \infty} \left[sf(s) \right] \tag{2.32}$$

Example 2.5

Prove by using the following equation that the left-hand side of Equation (2.32) is equal to its right-hand side:

$$f(t) = \frac{\mu}{(\lambda + \mu)} + \frac{\lambda}{(\lambda + \mu)} e^{-(\lambda+\mu)t} \tag{2.33}$$

where
  λ and μ are constants.

By inserting Equation (2.33) into the left-hand side of Equation (2.32), we obtain

$$\lim_{t \to \infty} \left[\frac{\mu}{(\lambda + \mu)} + \frac{\lambda}{(\lambda + \mu)} e^{-(\lambda+\mu)t} \right] = \frac{\mu}{(\lambda + \mu)} \tag{2.34}$$

By using Table 2.1 and Equation (2.33), we obtain

$$f(s) = \frac{\mu}{s(\lambda + \mu)} + \frac{\lambda}{(\lambda + \mu)} \cdot \frac{1}{(s + \lambda + \mu)} \tag{2.35}$$

By inserting Equation (2.35) into the right-hand side of Equation (2.32), we obtain

$$\lim_{s \to 0} \left[\frac{s\mu}{s(\lambda + \mu)} + \frac{s\lambda}{(\lambda + \mu)} \cdot \frac{1}{(s + \lambda + \mu)} \right] = \frac{\mu}{(\lambda + \mu)} \tag{2.36}$$

The right-hand sides of Equations (2.34) and (2.36) are the same. Thus, it proves that the left-hand side of Equation (2.32) is equal to its right-hand side.

2.6 PROBABILITY DISTRIBUTIONS

This section presents a number of probability distributions considered useful to perform various types of studies in the area of system safety, maintainability, and maintenance [10].

2.6.1 BINOMIAL DISTRIBUTION

This discrete random variable distribution is used in circumstances where one is concerned with the probabilities of outcome such as the number of occurrences (e.g., failures) in a sequence of n trials. More clearly, each trial has two possible outcomes (e.g., success or failure), but the probability of each trial remains constant or unchanged.

It is to be noted that this distribution is also known as the Bernoulli distribution, after Jakob Bernoulli (1654–1705) its founder [1]. The binomial probability density function, $f(y)$, is expressed by

$$f(y) = \binom{n}{i} p^y q^{n-y}, \text{ for } y = 0, 1, 2, \dots, n \tag{2.37}$$

where

$$\binom{n}{i} = \frac{n!}{i!(n-i)!}$$

y is the number of non-occurrences (e.g., failures) in n trials.
p is the single trial probability of occurrence (e.g., success).
q is the single trial probability of non-occurrence (e.g., failure).

The cumulative distribution function is given by

$$F(y) = \sum_{i=0}^{y} \binom{n}{i} p^i q^{n-i} \tag{2.38}$$

where

> $F(y)$ is the cumulative distribution function or the probability of y or fewer non-occurrences (e.g., failures) in n trials.

Using Equations (2.26) and (2.37), the expected value or the mean of the distribution is

$$E(y) = np \qquad (2.39)$$

2.6.2 NORMAL DISTRIBUTION

The normal distribution is a widely used continuous random variable distribution, and sometimes, it is called the Gaussian distribution after its founder Carl Friedrich Gauss (1777–1855), a German mathematician. The probability density function of the distribution is defined by

$$f(t) = \frac{1}{\sigma\sqrt{2\pi}} \exp\left[-\frac{(t-\mu)^2}{2\sigma^2}\right], -\infty < t < +\infty \qquad (2.40)$$

where

> μ and σ are the distribution parameters (i.e., mean and standard deviation, respectively).

Using Equations (2.22) and (2.40), we get the following cumulative distribution function:

$$F(t) = \frac{1}{\sigma\sqrt{2\pi}} \int_{-\infty}^{t} \exp\left[-\frac{(t-\mu)^2}{2\sigma^2}\right] dx \qquad (2.41)$$

Inserting Equation (2.40) into Equation (2.25) yields the following equation for the distribution mean value:

$$E(t) = m = \frac{1}{\sigma\sqrt{2\pi}} \int_{-\infty}^{\infty} t \exp\left[-\frac{(t-\mu)^2}{2\sigma^2}\right] dt \qquad (2.42)$$

$$= \mu$$

2.6.3 EXPONENTIAL DISTRIBUTION

The exponential distribution is a continuous random variable distribution that is widely used in the industrial sector, in performing reliability, safety, maintainability, and maintenance studies [11]. The distribution's probability density function is defined by

$$f(t) = \lambda e^{-\lambda t}, t \geq 0, \lambda > 0 \tag{2.43}$$

where

$f(t)$ is the probability density function.
t is time.
λ is the distribution parameter.

By substituting Equation (2.43) into Equation (2.22), we obtain the following expression for the cumulative distribution function:

$$F(t) = 1 - e^{-\lambda t} \tag{2.44}$$

Using Equations (2.43) and (2.25), we get the following expression for the distribution mean value:

$$E(t) = m = \frac{1}{\lambda} \tag{2.45}$$

where

m is the mean value.

2.6.4 RAYLEIGH DISTRIBUTION

This continuous random variable distribution is named after John Rayleigh (1842–1919), its founder [1]. The probability density function of the distribution is expressed by

$$f(t) = \left(\frac{2}{\alpha^2}\right) t e^{-\left(\frac{t}{\alpha}\right)^2}, t \geq 0, \alpha > 0 \tag{2.46}$$

where

α is the distribution parameter.

By inserting Equation (2.46) into Equation (2.22), we get the following cumulative distribution function:

$$F(t) = 1 - e^{-\left(\frac{t}{\alpha}\right)^2} \tag{2.47}$$

Using Equation (2.46) and Equation (2.25), we obtain the following expression for the distribution mean value:

$$E(t) = m = \alpha \Gamma\left(\frac{3}{2}\right) \tag{2.48}$$

where

$\Gamma(.)$ is the gamma function, which is defined by

$$F(y) = \int_0^\infty t^{y-1} e^{-t} dt, \text{for } y > 0 \tag{2.49}$$

2.6.5 WEIBULL DISTRIBUTION

This continuous random variable distribution is named after W. Weibull, a Swedish mechanical engineering professor, who developed it in the early 1950s [12]. The distribution can be used for representing many different physical phenomena, and its probability density function is expressed by

$$f(t) = \frac{ct^{c-1}}{\lambda^c} e^{-\left(\frac{t}{\lambda}\right)^c}, t \geq 0, c > 0, \lambda > 0 \tag{2.50}$$

where
λ and c are the distribution scale and shape parameters, respectively.
By inserting Equation (2.50) into Equation (2.22), we obtain the following cumulative distribution function:

$$F(t) = 1 - e^{-\left(\frac{t}{\lambda}\right)^c} \tag{2.51}$$

It is to be noted that exponential and Rayleigh distributions are the special cases of this distribution for $c = 1$ and $c = 2$, respectively.

Using Equation (2.50) and Equation (2.25), we get the following equation for the distribution mean value:

$$E(t) = m = \lambda \Gamma = \left(1 + \frac{1}{c}\right) \tag{2.52}$$

2.6.6 GAMMA DISTRIBUTION

This is a two-parameter distribution, and in 1961, it was considered as a possible model in life test-related problems [13]. The distribution probability density function is expressed by

$$f(t) = \frac{\mu(\mu t)^{N-1}}{\Gamma(N)} \exp(-\mu t), \quad t \geq 0, \mu > 0, N > 0 \tag{2.53}$$

where

t is time.

N is the shape parameter.

$\Gamma(N)$ is the gamma function.

$\mu = \dfrac{1}{\alpha}, \alpha$ is the scale parameter.

Inserting Equation (2.53) into Equation (2.22), we obtain the following equation for the cumulative distribution function:

$$F(t) = 1 - \frac{\Gamma(N, \mu t)}{\Gamma(N)} \tag{2.54}$$

where

$\Gamma(N, \mu t)$ is the incomplete gamma function.

Inserting Equation (2.53) into Equation (2.25), we obtain the following equation for the distribution mean:

$$E(t) = \frac{N}{\mu} \tag{2.55}$$

Finally, it is to be noted that three special case distributions of the gamma distribution are the exponential distribution, the chi-square distribution, and the special case Erlangian distribution [14].

2.7 SOLVING FIRST-ORDER DIFFERENTIAL EQUATIONS USING LAPLACE TRANSFORMS

Often, Laplace transforms are used for finding solutions to linear first-order differential equations, particularly when a set of linear first-order differential equations is involved. The following example demonstrates the finding of solutions to a set of differential equations describing an engineering system.

Example 2.6

Assume that an engineering system can be in either of the three states: operating normally, failed safely, or failed unsafely. The following three differential equations describe each of these engineering system states:

$$\frac{dP_0(t)}{dt} + (\lambda_1 + \lambda_2) P_0(t) = 0 \tag{2.56}$$

$$\frac{dP_1(t)}{dt} - \lambda_1 P_0(t) = 0 \tag{2.57}$$

$$\frac{dP_0(t)}{dt} - \lambda_2 P_0(t) = 0 \tag{2.58}$$

where

$P_i(t)$ is the probability that the engineering system is in state i at time t, for
$i = 0$ (operating normally), $i = 1$ (failed unsafely), and $i = 2$ (failed safely).
λ_2 is the engineering system constant safe failure rate.
λ_1 is the engineering system constant unsafe failure rate.

At time $t = 0$, $P_0(0) = 1$, $P_1(0) = 0$, and $P_2(0) = 0$.
Solve Equations (2.56)–(2.58) by using Laplace transforms.
By using Table 2.1, the specified initial conditions, and Equations (2.56)–(2.58),
we obtain

$$sP_0(s) + (\lambda_1 + \lambda_2) P_0(s) = 1 \tag{2.59}$$

$$sP_1(s) - \lambda_1 P_0(s) = 0 \tag{2.60}$$

$$sP_2(s) - \lambda_2 P_0(s) = 0 \tag{2.61}$$

Solving Equations (2.59)–(2.61), we get

$$P_0(s) = \frac{1}{(s + \lambda_1 + \lambda_2)} \tag{2.62}$$

$$P_1(s) = \frac{\lambda_1}{s(s + \lambda_1 + \lambda_2)} \tag{2.63}$$

$$P_2(s) = \frac{\lambda_2}{s(s + \lambda_1 + \lambda_2)} \tag{2.64}$$

Taking the inverse Laplace transforms of Equations (2.62)–(2.64), we obtain

$$P_0(t) = e^{-(\lambda_1 + \lambda_2)t} \tag{2.65}$$

$$P_1(t) = \frac{\lambda_1}{(\lambda_1 + \lambda_2)} \left[1 - e^{-(\lambda_1 + \lambda_2)t} \right] \tag{2.66}$$

$$P_2(t) = \frac{\lambda_2}{(\lambda_1 + \lambda_2)} \left[1 - e^{-(\lambda_1 + \lambda_2)t} \right] \tag{2.67}$$

Thus, Equations (2.65)–(2.67) are the solutions to Equations (2.56)–(2.58).

PROBLEMS

1. Write an essay on the historical developments in the area of mathematics.
2. Assume that the maintenance department of an organization inspected six identical systems and found 6, 10, 2, 4, 8, and 12 defects in each system. Calculate the average number of defects per system (i.e., arithmetic mean).
3. Define the following term:
   - Standard deviation
4. Describe the following two terms:
   - Commutative law
   - Distributive law
5. Define probability and write down five probability properties.
6. Define the following three terms:
   - Laplace transform
   - Cumulative distribution function
   - Expected value
7. Define the probability density of the gamma distribution.
8. What are the special case distributions of the Weibull distribution?
9. Prove that the sum of Equations (2.65)–(2.67) is equal to unity.
10. Prove that the mean of the Rayleigh distribution is given by Equation (2.48).

REFERENCES

1. Eves, H., *An Introduction to the History of Mathematics*, Holt, Rinehart and Winston, New York, 1976.
2. Owen, D.B., Ed., *On the History of Statistics and Probability*, Marcel Dekker, New York, 1976.
3. Lipschutz, S., *Set Theory*, McGraw Hill Book Company, New York, 1964.
4. Fault Tree Handbook, Report No. NUREG-0492, U.S. Nuclear Regulatory Commission, Washington, D.C., 1981.
5. Mann, N.R., Schefer, R.E., Singpurwalla, N.D., *Methods for Statistical Analysis of Reliability and Life Data*, John Wiley & Sons, New York, 1974.
6. Lipschutz, S., *Probability*, McGraw-Hill, New York, 1965.
7. Shooman, M.L., *Probabilistic Reliability: An Engineering Approach*, McGraw-Hill, New York, 1968.
8. Spiegel, M.R., *Laplace Transforms*, McGraw-Hill, New York, 1965.
9. Oberhettinger, F., Badic, L., *Table of Laplace Transforms*, Springer-Verlag, New York, 1973.
10. Patel, J.K., Kapadia, C.H., Owen, D.B., *Handbook of Statistical Distributions*, Marcel Dekker, New York, 1976.
11. Davis, D.J., An Analysis of Some Failure Data, *Journal of the American Statistical Association*, Vol. 48, June 1952, pp. 113–150.
12. Weibulll, W., A Statistical Distribution Function of Wide Applicability, *Journal of Applied Mechanics*, Vol. 18, 1951, pp. 293–297.
13. Gupta, S., Groll, P., Gamma Distribution in Acceptance Sampling Based on Life Tests, *Journal of American Statistical Association*, Vol. 56, December 1961, pp. 942–970.
14. Dhillon, B.S., *Mechanical Reliability: Theory, Models and Applications*, American Institute of Aeronautics and Astronautics, Washington, D.C., 1988.

3 Introduction to Safety

3.1 NEED FOR SAFETY AND ENGINEERING SAFETY GOALS

The desire to be safe and secure has always been a very important concern to humans. For example, early humans took appropriate precautions for safeguarding against natural hazards around them. Moreover, in 2000 B.C., Hammurabi, an ancient Babylonian ruler, developed a code referred to as the Code of Hammurabi. The code included clauses on items such as monetary damages against people who caused injury to others and allowable fees for involved physicians [1, 2].

Nowadays, safety has become a very important issue because each year, a large number of people die and get seriously injured due to workplace-related and other accidents. For example, in 1996, in the United States as per the National Safety Council (NSC), there were 93,400 deaths and a vast number of disabling injuries due to accidents [3]. The total cost of these accidents was estimated to be approximately $121 billion. Some of the other factors that are also playing an important role in demanding the need for better safety are increasing number of law suits, government regulations, and public pressures.

There are many goals of engineering safety. Some of these goals are as follows [4]:

- Eliminate or control hazards.
- Develop new techniques and methods to improve safety.
- Maximize public confidence in regard to product safety.
- Reduce accidents.
- Maximize returns on safety-related efforts.

3.2 SAFETY-ASSOCIATED FACTS AND FIGURES

Some of the facts and figures directly or indirectly concerned with safety are as follows:

- In 1980, American employers spent approximately $22 billion to insure or self-insure against job-associated injuries [5].
- In 2000, there were approximately 97,300 unintentional injury deaths in the United States. Their cost was estimated to be about $512.4 billion [6].
- In 2000, there were 3.8 deaths per 100,000 workers in the United States [7].
- In the European Union (EU), each year over 4.5 million accidents result in absence from work for more than three days [8]. This translates into a total loss of approximately 146 million working days [8].
- In 2000, a total of 5200 persons died due to work-associated accidents in the United States [6].

DOI: 10.1201/9781003365204-3

- In the EU, each year around 5500 people die due to work-associated accidents [8].
- In 2000, the total cost of work-associated injuries in the United States was approximately $131.2 billion [7].
- In the United States, in a typical year, approximately 35 million work hours are lost due to accidents [9].
- In 1997, three workers in the United States were awarded $5.8 million after they sued a computer equipment manufacturing company for musculoskeletal disorders (MSDs) [10].
- During the period 1960–2000, work-associated accidental deaths in the United States dropped by 60% (i.e., 13,800 in 1960 to 5200 in 2000) [7, 11].
- In 1995, the total cost of work-associated accidents in the United States was estimated to be approximately $75 billion [12].
- In the 1990s, the cost of accidents per worker per year in the United States was approximately $420 [2].

3.3 ENGINEERS AND SAFETY

In the modern times, problems with engineering products' safety may be traced back to railroads. For example, the very day Stephenson's first railroad line was dedicated, a fatal railroad accident took place (i.e., a prominent English legislator was killed) [10]. One year later, the boiler of the first locomotive built in the United States exploded and killed one person and badly injured a number of fuel servers [10, 13].

Today, engineering systems have become very complex and sophisticated. These systems' safety is a challenging issue, because of competition and other factors engineers are pressured to complete new designs quickly and at lower costs. Past experiences over the years clearly indicate that this, in turn, usually results in more design deficiencies, errors, and causes of accidents. The design deficiencies can cause or contribute to accidents. The design deficiency may take place because of a designer/ design [10]:

- Overlooked to warn adequately of a potential hazard.
- Overlooked to foresee an unexpected application of an item or its all potential consequences.
- Overlooked to prescribe a proper operational procedure in situations where a hazard might exist.
- Overlooked to reduce or eliminate the occurrence of human error.
- Overlooked to provide a satisfactory level of protection in a worker's/user's personal protective equipment.
- Is unfinished, incorrect, or confusing.
- Creates an arrangement of operating controls and other devices that substantially increases reaction time in emergency situations or is conducive to errors.
- Relies on product users to avoid an accident.
- Incorporates poor warning mechanisms instead of providing a rather safe design to eradicate hazards.

- Does not adequately determine or consider the failure, error, action, or omission consequences.
- Violates usual capabilities/tendencies of potential users.
- Creates an unsafe characteristic of an item.
- Places an unreasonable level of stress on operators.

3.4 PRODUCT HAZARD CATEGORIES AND COMMON MECHANICAL-RELATED INJURIES

There are many product-associated hazards. These hazards may be categorized under the following six categories [14]:

- **Environmental hazards:** These hazards may be grouped under two classifications: internal and external. The internal hazards are concerned with the changes in the surrounding environment that lead to internally damaged product. These hazards can be minimized or eliminated by carefully considering factors such as electromagnetic radiation, extremes of temperatures, vibrations, ambient noise levels, illumination level, and atmospheric contaminants during the design phase.

    The external hazards are the hazards posed by the product under consideration during its life span. These hazards include disposal hazards, services-life operation hazards, and maintenance hazards.
- **Energy hazards:** These hazards may be grouped under two classifications: kinetic energy and potential energy. The kinetic energy hazards pertain to parts that have energy because of their motion. Some examples of these parts are fan blades, flywheels, and loom shuttles. Any object that interferes with their motion can experience substantial damage.

    The potential energy pertains to parts that store energy. Such parts include springs, compressed-gas receivers, electronic capacitors, and counterbalancing weights. During the servicing of the equipment, such hazards are important because stored energy can cause quite serious injury when released suddenly.
- **Human factors hazards:** These hazards are associated with poor design in regard to people. More specifically, to their physical strength, visual angle, intelligence, weight, computational ability, education, visual acuity, height, length of reach, etc.
- **Kinematic hazards:** These hazards are associated with situations where parts come together while moving and lead to possible cutting, pinching, or crushing of any object caught between them.
- **Electrical hazards:** These hazards have two principal components: electrocution hazard and shock hazard. The major electrical hazard to product or property stems from electrical faults, often referred to as short circuits.
- **Misuse and abuse-associated hazards:** These hazards are associated with the product usage by people. Past experiences over the years clearly indicate that product misuse can result in serious injuries. Product abuse can also lead to hazardous situations or injuries and some of the causes for the abuse are lack of proper maintenance and poor operating practices.

In the industrial sector, humans interact with various types of equipment to conduct tasks such as drilling, cutting, punching, stitching, shaping, chipping, stamping, and abrading. There are various types of injuries that can result in conducting such tasks. Some of the common ones are as follows [2]:

- **Breaking-related injuries:** These injuries are generally associated with machines used for deforming various types of engineering materials. Often, break in a bone is called a fracture. In turn, fracture is classified into many categories including complete, incomplete, transverse, compound, comminuted, simple, and oblique.
- **Shearing-related injuries:** These injuries pertain to shearing processes. In manufacturing, power-driven shears are widely used for performing various types of tasks, including severing paper, plastic metal, and elastomers. In the past, in using such machines, tragedies such as amputation of hands/fingers have occurred.
- **Puncturing-related injuries:** These injuries occur when an object penetrates straight into a person's body and pulls straight out. In the industrial setting, normally, these types of injuries pertain to punching machines because they have sharp tools.
- **Crushing-related injuries:** These injuries occur where a body part is caught between two hard surfaces moving progressively together and crushing any object that comes between them.
- **Straining and spraining-related injuries:** In the industrial environment, there are many opportunities associated with the use of machines or other tasks, for the occurrence of straining-and spraining-related injuries, for example, spraining of ligaments or straining or muscles.
- **Cutting and tearing-related injuries:** These injuries occur when a person's body part comes in contact with a sharp edge. The severity of a tear or a cut depends upon the degree of damage to items such as skin, veins, muscles, and arteries.

3.5 PRODUCT LIABILITY EXPOSURE CAUSES AND WORKER'S COMPENSATION

Nowadays, product liability is an important factor in safety. It is to be noted that until 1960, in the United States, manufacturers were not liable unless they manufactured flagrantly dangerous products/items. However, today manufacturers are increasingly being sued by the product users, misusers, and even by abusers.

Past experiences over the years clearly indicate that about 60% of the liability-related cases involved failure to provide adequate danger warning labels on manufactured products. Nonetheless, some of the common causes of product liability exposure are as follows [15]:

- Inadequate research during product development.
- Inadequate testing of product prototypes.
- Faulty product design.
- Poorly written warnings.

- Faulty manufacturing.
- Poorly written instructions.

Due to various government legislations, workers' compensation has become a very important factor in workplace safety. The workers' compensation law was first passed in 1838 in Prussia for protecting railroad workers [1]. However, in the United States, it was only in 1908 when the first workers' compensation law was passed for protecting federal government workers conducting various types of hazardous tasks. In the current work environment, in general, all American workers are well protected by compensation laws.

Nonetheless, a study of Workers' Compensation Laws was carried out by the Unites States Chamber of Commerce. It concluded their following seven underlying objectives [4, 10]:

- To encourage frank and fair study of accidents' causes.
- To minimize, as much as possible, human suffering and preventable accidents.
- To free charities from finance-associated burdens created by uncompensated workplace accidents.
- To provide an appropriate approach for reducing the degree of personal injury litigation in courts.
- To eliminate time-consuming and costly trials and appeals.
- To maximize employer involvement in safety and rehabilitation through the use of an experience-rating mechanism.
- To provide fast and reasonable income and medical benefits to victims of work-accidents or income benefits to dependents of victims, irrespective faults.

Additional information on workers' compensation is available in Refs. [2, 10, 15–17].

PROBLEMS

1. Discuss the need for safety.
2. What are the principal goals of safety?
3. List at least ten safety-associated facts and figures.
4. Discuss design deficiency factors with respect to designer/design.
5. Discuss six categories of product hazards.
6. Discuss common mechanical-related injuries.
7. What are the common causes of product liability exposure?
8. Write an essay on workers' compensation.
9. Discuss the following terms:
   - Internal environmental hazards
   - External environmental hazards
   - Kinetic energy
   - Potential energy
10. Compare environmental hazards with energy hazards in regard to product hazards.

REFERENCES

1. Ladon, J., Ed., *Introduction to Occupational Health and Safety*, National Safety Council (NSC), Chicago, IL, 1986.
2. Goetsch, D.L., *Occupational Safety and Health*, Prentice Hall, Englewood Cliffs, New York, 1996.
3. Accidental Facts, Report, National Safety Council, Chicago, IL, 1996.
4. Dhillon, B.S., *Engineering Safety: Fundamentals, Techniques and Applications*, World Scientific Publishing, River Edge, New Jersey, 2003.
5. Lancianese, F., The Soaring Costs of Industrial Accidents, *Occupational Hazards*, August 1983, pp. 30–35.
6. Report on Injuries in America in 2000, National Safety Council, Chicago, IL, 2000.
7. Report on Injuries in America, National Safety Council, Chicago, IL, 2001.
8. How to Reduce Workplace Accidents, *European Agency for Safety and Health at Work*, Belgium, Brussels, 2001.
9. Accident Facts, National Safety Council, Chicago, IL, 1990–1993.
10. Hammer, W., Price, D., *Occupational Safety Management and Engineering*, Prentice Hall, Upper Saddle River, New Jersey, 2001.
11. Blake, R.P., Ed., *Industrial Safety*, Prentice Hall, Englewood Cliffs, New Jersey, 1964.
12. Spellman, F.R., Whitting, N.E., *Safety Engineering: Principles and Practice*, Government Institutes, Rockville, MD, 1999.
13. Operator Safety, Engineering, May 1974, pp. 358–363.
14. Hunter, T.A., *Engineering Design for Safety*, McGraw-Hill, New York, 1992.
15. State Workers' Compensation Laws, The U.S. Department of Labour, Washington, D.C., January 1990.
16. Somers, A.R., Somers, H.M., *Women's Compensation*, John Wiley and Sons, New York, 1945.
17. Eastman, C., *Work Accidents and the Law*, New York Charities Publication Committee, New York, 1910.

4 Safety Analysis Methods

4.1 INTRODUCTION

Although the problem of safety has been around for a long time, the development of safety analysis methods is relatively new. Some of these methods were specifically developed for application in the area of safety and the others for application in different areas. Some examples of these methods are failure modes and effect analysis (FMEA), fault tree analysis (FTA), and quality control charts. FTA and FMEA were developed for application in reliability areas and the quality control charts for use in quality control work. Nonetheless, the main objective of all these safety analysis methods is to stop the occurrence of accidents and hazards.

As the effectiveness of these methods can vary considerably from one application to another, a careful consideration is absolutely necessary in selecting an appropriate method for a specific application. This chapter presents a number of important safety analysis methods [1–5].

4.2 HAZARDS AND OPERABILITY ANALYSIS (HAZOP)

This method was developed for use in chemical industry and is extremely useful for highlighting safety-associated problems prior to availability of full data concerning an item [6, 7]. Three fundamental objectives of the HAZOP are as follows [1–5]:

- Produce a complete description of process/facility.
- Review each process/facility element to determine how deviations from the design intentions can happen.
- Decide whether the deviations can lead to operating problems/hazards.

A HAZOP study is conducted by following seven steps presented below [3, 8]:

- **Step I:** Choose the process/system to be analyzed.
- **Step II:** Establish the team of appropriate experts.
- **Step III:** Describe the HAZOP process to all persons forming the team.
- **Step IV:** Establish goals and time schedules.
- **Step V:** Conduct brainstorming sessions as considered appropriate.
- **Step VI:** Conduct analysis.
- **Step VII:** Document the study.

It is to be noted that HAZOP has basically the same weaknesses as FMEA. For example, they both predict problems related to system/process failures, but do not factor human error into the equation. This is the key weakness because human error is often a factor in accidents.

DOI: 10.1201/9781003365204-4

4.3   FAULT TREE ANALYSIS (FTA)

This method was developed in the early 1960s at the Bell Laboratories for performing reliability and safety analyses of the Minuteman Launch Control System. Today,
it is used widely in industry to perform various types of reliability and safety analyses. FTA starts by defining a system's undesirable state (event) and then analyzes
the system to determine all possible situations that can lead to the occurrence of an
undesirable event. Thus, it highlights all possible failure causes at all possible levels
related with a system under consideration and the relationship between causes. FTA
can be used to analyze various types of safety problems.

The method (i.e., FTA) uses various types of symbols, and four commonly used
symbols in fault tree construction are shown in Figure 4.1 [9].

The circle denotes a basic fault event or the failure of an elementary component/
part. The event's probability of occurrence and failure and repair rates are generally
obtained from empirical data. The rectangle denotes a fault event that results from
the combination of fault events through the input of a logic gate. The AND gate
denotes that an output fault event occurs only if all the input fault events occur.
Finally, the OR gate denotes that an output fault event occurs if one or more of the
input fault events occur.

The probabilities of the occurrence of the output fault events of logic gates OR
and AND are given by

- **OR gate**

$$P(E_0) = 1 - \prod_{j=1}^{n} \{1 - P(E_j)\} \tag{4.1}$$

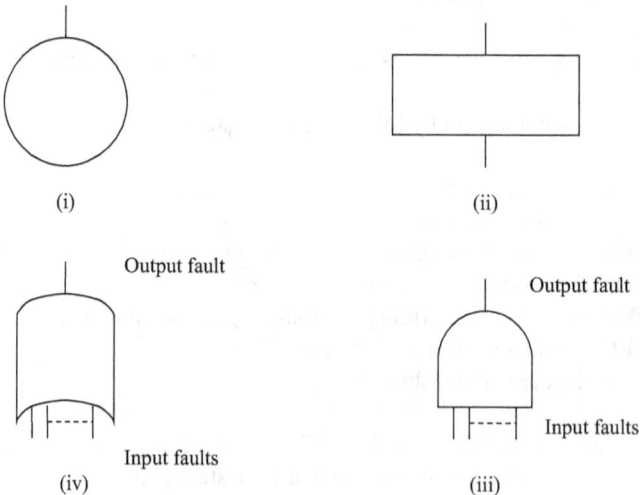

(i)

(ii)

Output fault

Output fault

(iv)

Input faults

(iii)

Input faults

FIGURE 4.1   Four commonly used fault tree symbols: (i) circle, (ii) rectangle, (iii) AND
gate, and (iv) OR gate.

where

$P(E_0)$ is the probability of occurrence of the OR gate output fault event E_0.
$P(E_j)$ is the probability of occurrence of input fault event E_j, for $j = 1,2,3,\ldots, n$.
n is the number of independent input fault events.

- **AND gate**

$$P(X_0) = \prod_{j=1}^{n} P(X_j) \tag{4.2}$$

where

$P(X_0)$ is the probability of occurrence of the AND gate output fault event X_0.
$P(X_j)$ is the probability of occurrence of the AND gate input fault event X_j, for $j = 1, 2, 3, \ldots, n$.

The application of FTA to a safety problem is demonstrated through the following example:

Example 4.1

Assume that an engineering system is located in a windowless room and is being operated by a human. For its safe operation by the human operator, the room must be lit all the time. The room has two light bulbs and a switch. The switch can only fail to close and the room is supplied with regular electrical power only. The probability of performing an unsafe operation by the operator increases quite dramatically, if there is total darkness in the room.

Develop a fault tree for the top event: dark room (i.e., unsafe system operation by the operator likely).

Using the symbols shown in Figure 4.1, a fault tree for the example is shown in Figure 4.2. It is to be noted that the single capital letters in Figure 4.2 diagram denote corresponding fault events [e.g., T: Dark room (top fault event)].

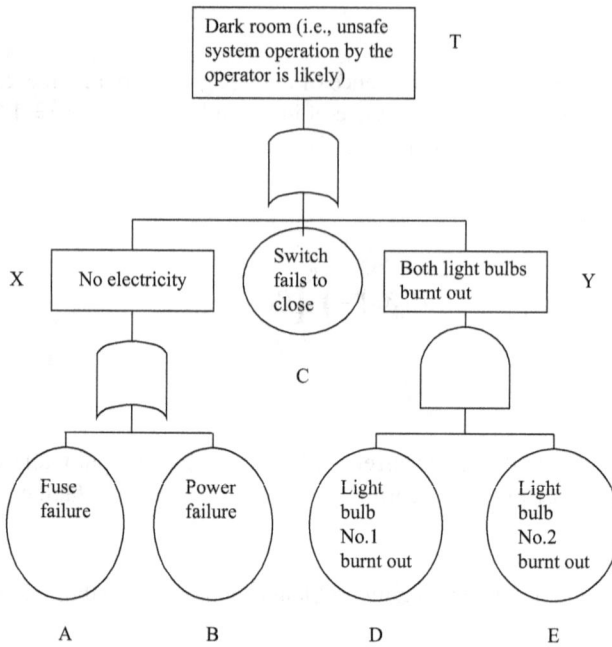

FIGURE 4.2 A fault tree for Example 4.1.

Example 4.2

Assume that the probabilities of occurrence of independent fault events A, B, C, D, and E in Figure 4.2 are 0.04, 0.05, 0.06, 0.07, and 0.08, respectively. Calculate the probability of occurrence of the top fault event T by using Equations (4.1) and (4.2).

By inserting the specified occurrence probability values of fault events A and B into Equation (4.1), we obtain

$$P(X) = 1 - \left[(1 - 0.04)(1 - 0.05)\right] = 0.088$$

where
P(X) is the probability of occurrence of fault event X (no electricity).

By substituting the given occurrence probability values of fault events D and E into Equation (4.2), we obtain

$$P(Y) = (0.07)(0.08) = 0.0056$$

where
P(Y) is the probability of occurrence of fault event Y (both light bulbs burnt out).

By inserting the given data value and the above two calculated values into Equation (4.1), we obtain

$$P(T) = 1 - \left[(1-0.06)(1-0.088)(1-0.0056)\right] = 0.8524$$

where
P(T) is the probability of occurrence of the top fault event T [dark room (i.e., unsafe system operation by the operator is likely)].

Thus, probability of occurrence of the top fault event T [dark room (i.e., unsafe system operation by the operator is likely)] is 0.8524. Figure 4.3 shows Figure 4.2 fault tree with the above calculated values and the specified fault event occurrence probability values.

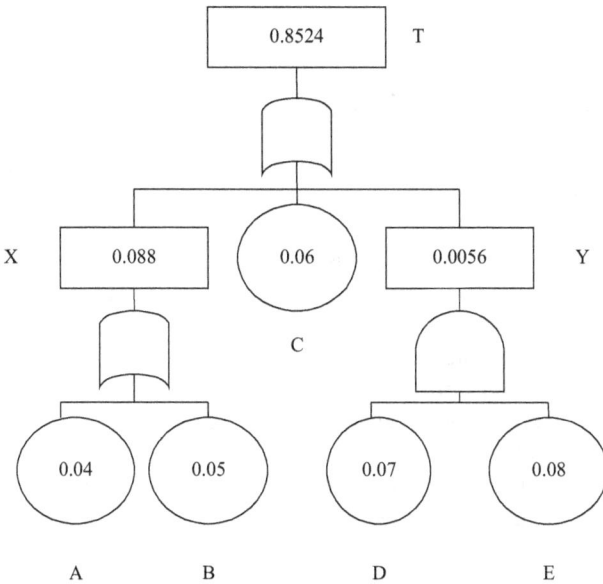

FIGURE 4.3   A fault tree with the given and calculated fault event occurrence probability values.

4.4   TECHNIC OF OPERATIONS REVIEW (TOR)

This method was developed by D.A. Weaver of the American Society of Safety Engineers (ASSE) in the early 1970s [3]. TOR may simply be described as a hands on analytical methodology/approach developed to determine the root system causes of an operation failure/malfunction. It uses a worksheet containing simple and straight-forward terms requiring yes/no decisions. The basis for the activation of the method is an incident occurring at a certain time and place involving certain individuals. The following eight steps are associated with the method [3, 16]:

- **Step I:** Form the TOR team.
- **Step II:** Hold a roundtable session with all involved team members. The main objective of this session is to communicate common knowledge to all involved parties.

- **Step III:** Highlight a single important factor that has played a pivotal role in the occurrence of accident/incident. This factor is the result of team consensus and serves as a starting point to all further investigations.
- **Step IV:** Use the team consensus in responding to a sequence of yes/no options.
- **Step V:** Evaluate all highlighted factors and ensure the existence of consensus among the members of the team.
- **Step VI:** Prioritize all the contributing factors under consideration.
- **Step VII:** Develop appropriate corrective/preventive strategies with respect to each and every contributing factor under consideration.
- **Step VIII:** Implement strategies.

Finally, it is to be noted that the strength of this method stems from the involvement of line personnel in the analysis and its weakness from the fact that it is specifically designed as an after-the-fact-process.

4.5   FAILURE MODES AND EFFECT ANALYSIS (FMEA)

This is a widely used method for performing reliability analysis of engineering systems. It analyzes each potential failure mode in a given system for determining the effects of such failure modes on the entire system. The method can also be used to perform various types of safety analysis.

FMEA is essentially composed of the following seven main steps [11]:

- **Step I:** Establish system definition (i.e., define the system and its all associated requirements).
- **Step II:** Establish ground rules for conducting FMEA.
- **Step III:** Develop system hardware description.
- **Step IV:** Develop description of system functional blocks.
- **Step V:** Highlight failure modes and their effects for each and every item in the system.
- **Step VI:** Compile critical item list.
- **Step VII:** Document the analysis.

Additional information on this method is available in Ref. [12].

4.6   JOB SAFETY ANALYSIS (JSA)

This method is concerned with discovering and rectifying potential hazards that are intrinsic to or inherent in a given workplace. Generally, personnel such as worker, supervisor, and safety professional participate in job safety analysis. The following five steps are associated with the performance of job safety analysis [13]:

- **Step I:** Select a job for analysis.
- **Step II:** Break down the job into various tasks/steps.
- **Step III:** Highlight all possible hazards and propose suitable measures for effectively controlling them to appropriate levels.

- **Step IV:** Apply the proposed measures.
- **Step V:** Evaluate the final results.

Past experiences over the years clearly indicate that the success of job safety analysis depends on the degree of rigor exercised by the job safety analysis team during the analysis process.

4.7   CONTROL CHARTS

There are many types of control charts and they were originally developed by Walter A. Shewhart in 1924 for use in quality control-related work [14]. These charts can also be used to perform various types of safety analysis. A control chart is a graphical technique used for evaluating whether a given process is in a "state of statistical control" or out of control. More specifically, when a sample value falls outside the upper and lower control limits of a control chart, it means that the process is out of statistical control and needs an investigation.

In safety work, the process could be the accidents' frequency, accidents' severity, etc. There are many types of control charts that can be utilized in safety-related studies. Here, their application to safety-associated problems is demonstrated through one type of control chart only (i.e., the C-chart). The C-chart is based on the Poisson distribution. Poisson distribution's mean and standard deviation are defined by [14].

$$\mu = \frac{\theta}{T} \tag{4.3}$$

where

μ is the mean of the Poisson distribution.
θ is the total number of accidents.
T is the total time period.

and

$$\sigma = \left(\mu\right)^{1/2} \tag{4.4}$$

where

σ is the standard deviation of the Poisson distribution.

Thus, the lower and upper control limits of the C-chart are defined by [14]

$$LCL_{cc} = \mu - 3\sigma \tag{4.5}$$

and

$$UCL_{cc} = \mu + 3\sigma \tag{4.6}$$

where

LCL_{cc} is the lower control limit of the C-chart.
UCL_{cc} is the upper control limit of the C-chart.

LCL_{cc} is the lower control limit of the C-chart.
UCL_{cc} is the upper control limit of the C-chart.

The following example demonstrates the application of the C-chart to a safety-related problem.

Example 4.3

Assume that an organization operating various types of engineering systems reported a total of 165 accidents over a period of ten years. The breakdowns of these accidents for year 1, 2, 3, 4, 5, 6, 7, 8, 9, and 10 were 15, 10, 20, 20, 10, 15, 20, 5, 15, 20, and 15, respectively.

Calculate the values of μ, σ, and the lower and upper control limits of the C-chart and comment on the end results.

By inserting the given data into Equation (4.3), we obtain

$$\mu = \frac{165}{10} = 16.5 \text{ accidents/year}$$

Thus, from Equation (4.4), we get

$$\sigma = (16.5)^{1/2} = 4.06$$

By substituting the above-calculated values into Equations (4.5) and (4.6), we obtain

$$LCL_{cc} = 16.5 - 3(4.06) = 4.32$$

and

$$UCL_{cc} = 16.5 + 3(4.06) = 28.68$$

Thus, the values of μ, σ, and the lower and upper control limits of the C-chart are 16.5, 4.06, 4.32, and 28.68, respectively. It means that the occurrences of accidents are within the lower and upper control limits of the C-chart.

4.8   INTERFACE SAFETY ANALYSIS (ISA)

Interface Safety Analysis (ISA) is concerned with determining the incompatibilities between subsystems and assemblies of a given product/item that may result in accidents. More specifically, ISA establishes that distinct units/parts/items can be integrated into a viable system and an individual part's/unit's normal operation will not impair the performance or cause damage to another unit/part or the overall system. Although this method considers various relationships, they can be grouped under the following three classifications [15]:

- **Classification I: Flow relationships.** These relationships are associated with two or more units. Moreover, the flow between two units may involve items such as electrical energy, fuel, air, lubricating oil, water, or steam. Usually, the common problems associated with many products are the effective flow of fluids and energy from one item to another through confined passages, thus resulting in direct/indirect safety-associated problems. All in all, in regard to fluids, factors such as loss of pressure, contamination, lubricity, corrosiveness, flammability, and toxicity must be considered with care from the safety aspect.
- **Classification II: Physical relationships.** These relationships are associated with the physical aspects of units/items. For example, some units/items may operate quite well individually, but they may fail to fit together due to dimensional differences or other problems.
- **Classification III: Functional relationships.** These relationships are associated with multiple units/items. For example, in a situation where outputs of an item/unit constitute the inputs to a downstream unit/item, any error in outputs and inputs may cause damage to the downstream unit/item and in turn a safety hazard.

4.9   MARKOV METHOD

This is one of the most widely used methods to perform various types of safety, reliability, and maintenance analysis and is named after a Russian mathematician Andrei A. Markov (1856–1922). The method is subject to the following three assumptions [16]:

- The transitional probability from one state to another in the finite time interval Δt is given by $\lambda \Delta t$, where λ is the constant transition rate (e.g., the failure or repair rate) from one system state to another.
- The probability of occurrence of more than one transition in finite time interval Δt from one state to another is very small or negligible (e.g., $[\lambda \Delta t]$ $[\lambda \Delta t] \rightarrow 0$).
- All occurrences are independent of each other.

The following example demonstrates the application of this method to a safety problem.

Example 4.4

Assume that an engineering system can either fail safely or unsafely. The safe and unsafe failure rates of an engineering system are constant. The state space diagram of the engineering system is shown in Figure 4.4. The numerals in boxes denote the engineering system states. Develop equations for the engineering system state probabilities. Assume that all engineering system failures are statistically independent.

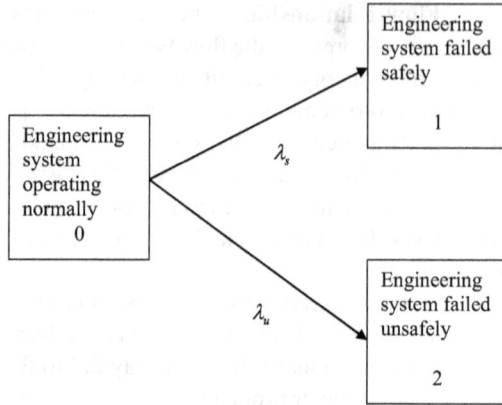

FIGURE 4.4   The engineering state space diagram.

With the aid of the Markov method, we write down the following Equations for Figure 4.4 diagram [2]:

$$\frac{dP_0(t)}{dt} + P_0(t)\lambda_s + P_0(t)\lambda_u = 0 \tag{4.7}$$

$$\frac{dP_1(t)}{dt} - P_0(t)\lambda_s = 0 \tag{4.8}$$

$$\frac{dP_2(t)}{dt} - P_0(t)\lambda_u = 0 \tag{4.9}$$

At time $t = 0$, $P_0(0) = 1$, $P_1(0) = 0$, and $P_2(0) = 0$. The symbols used in Equations (4.7)–(4.9) are defined below.

$P_i(t)$ is the probability that the engineering system is in state i at time t, for $i = 0$ (operating normally), $i = 1$ (failed safely), and $i = 2$ (failed unsafely).
λ_s is the engineering system constant safe failure rate.
λ_u is the engineering system constant unsafe failure rate.

By solving Equations (4.7)–(4.9), we get

$$P_0(t) = e^{-(\lambda_s + \lambda_u)t} \tag{4.10}$$

$$P_1(t) = \frac{\lambda_s}{(\lambda_s + \lambda_u)}\left[1 - e^{-(\lambda_s + \lambda_u)t}\right] \tag{4.11}$$

$$P_2(t) = \frac{\lambda_u}{(\lambda_s + \lambda_u)}\left[1 - e^{-(\lambda_s + \lambda_u)t}\right] \tag{4.12}$$

Example 4.5

Calculate the probability that during 600 hours of operation, the engineering system will fail unsafely, if the engineering system safe and unsafe failure rates are 0.008 failures/hour and 0.002 failures/hour, respectively. Assume that all engineering system failures are statistically independent.

By substituting the given data values into Equation (4.12), we obtain

$$P_2(600) = \frac{0.002}{(0.008 + 0.002)} \left[1 - e^{-(0.008 + 0.002)(600)} \right]$$

$$= 0.1995$$

Thus, the probability that during 600 hours of operation, the engineering system will fail unsafely is 0.1995.

4.10   SAFETY INDEXES

These indexes are concerned with measuring an organization's safety performance. Over the years, many such indexes have been developed and two of these indexes proposed by the American National Standards Institute (ANSI) are presented below [17, 18].

4.10.1   DISABLING INJURY FREQUENCY RATE (DIFR)

This index is based on four events that occur during the time period covered by the rate (i.e., temporary disabilities, permanent partial disabilities, permanent disabilities, and deaths). The index is defined by

$$DIFR = \frac{DI_t(1,000,000)}{T_{ee}} \qquad (4.13)$$

where

DI_t is the total number of disabling injuries.
T_{ee} is the employee exposure time expressed in hours.

One important advantage of this index is that it considers differences in quantity of exposures due to varying employee/worker work hours, either within the framework of the organization during successive period or among organizations classified under the similar industry group [17].

4.10.2   DISABLING INJURY SEVERITY RATE (DISR)

This index is based on four factors occurring during the period covered by the rate (i.e., total scheduled charges (days) for all deaths, permanent total, and permanent partial disabilities, and the total number of days of disability from all temporary injuries).

The index is defined by

$$DISR = \frac{TDC(1,000,000)}{T_{ee}}$$ (4.14)

where

TDC is the total number of days charged.

Some of the benefits of the index are as follows:

- A quite useful tool to take into consideration differences in quantity of exposure over time.
- Useful for making a meaningful comparison between different organizations.
- A quite useful tool to answer the question: "How serious are injuries in our organization?"

PROBLEMS

1. Describe hazards and operability analysis (HAZOP).
2. Assume that an engineering system is located in a windowless room and is being operated by a human. For its safe operation by the human operator, the room must be lit all the time. The room has three light bulbs and a switch. The switch can only fail to close and the room is supplied with regular electrical power only. The probability of performing an unsafe operation by the operator increases quite dramatically, if there is total darkness in the room.

    Develop a fault tree for the top event: dark room (i.e., unsafe system operation by the operator likely).
3. List the steps used in performing technic of operations review (TOR).
4. Compare failure modes and effect analysis (FMEA) with fault tree analysis (FTA).
5. Describe job safety analysis (JSA).
6. Write an essay on the use of control charts in safety work.
7. Describe interface safety analysis (ISA).
8. Prove that in Example 4.4 the probability of the system failing unsafely is given by Equation (4.12).
9. Define the following two safety indexes:
    - Disabling injury frequency rate (DIFR)
    - Disabling injury severity rate (DISR)
10. What are the advantages of DISR and DIFR?

REFERENCES

1. Roland, H.E., Moriarty, B., *System Safety Engineering and Management*, John Wiley and Sons, New York, 1983.
2. Dhillon, B.S., *Engineering Safety: Fundamentals, Techniques, and Applications*, World Scientific Publishing, River Edge, New Jersey, 2003.
3. Goetsch, D.L., *Occupational Safety and Health*, Prentice Hall, Englewood Cliffs, New Jersey, 1996.
4. Gloss, D.S., Wardle, M.G., *Introduction to Safety Engineering*, John Wiley and Sons, New York, 1984.
5. Raouf, A., Dhillon, B.S., *Safety Assessment: A Quantitative Approach*, Lewis Publishers, Boca Raton, Florida, 1994.
6. Guidelines for Hazard Evaluation Procedures, American Institute of Chemical Engineers, New York, 1985.
7. Dhillon, B.S., Rayapati, S.N., Chemical Systems Reliability: A Survey, *IEEE Transactions on Reliability*, Vol. 37, 1988, pp. 199–208.
8. Risk Analysis Requirements and Guidelines, Report No. CAN/CSA-Q634-91, prepared by the Canadian Standards Association, 1991. Available from Canadian Standards Association, 178 Rexdale Boulevard, Rexdale, Ontario, Canada.
9. Dhillon, B.S., Singh, C., *Engineering Reliability: New Techniques and Applications*, John Wiley and Sons, New York, 1981.
10. Hallock, R.G., Technic of Operations Review Analysis: Determine Cause of Accident/Incident, *Safety and Health*, Vol. 60, August 1991, pp. 38–39, 46.
11. Jordan, W.E., Failure Modes, Effects, and Critical Analysis, *Proceedings of the Annual Reliability and Maintainability Symposium*, 1972, pp. 30–37.
12. Dhillon, B.S., *Design Reliability: Fundamentals and Applications*, CRC Press, Boca Raton, Florida, 1999.
13. Hammer, W., Price, D., *Occupational Safety Management and Engineering*, Prentice Hall, Upper Saddle River, New Jersey, 2001.
14. Statistical Quality Control Handbook, AT and T Technologies, Indianapolis, IN, 1956.
15. Hammer, W., *Product Safety Management and Engineering*, Prentice Hall, Englewood Cliffs, New Jersey, 1980.
16. Shooman, M.L., *Probabilistic Reliability: An Engineering Approach*, McGraw-Hill, New York, 1968.
17. Tarrants, W.E., *The Measurement of Safety Performance*, Garland STPM Press, New York, 1980.
18. Z-16.1, *Method of Recording and Measuring Work Injury Experience*, American National Standards Institute, New York, 1985.

5 Safety Management and Costing

5.1 INTRODUCTION

Today, safety management is a very important element of a safety program in an organization and its beginning appears to be the period during the 1950s and 1960s [1, 2]. In the safety management's history, an important milestone occurred in 1970 with the passage of the Occupational Safety and Health Act (OSHA) by the U.S. Congress. Since 1970, many new developments concerned with safety management have occurred and safety management may simply be defined as the accomplishment of safety through the efforts of others [1–6]. Nonetheless, safety management's fundamental objective is to eliminate human suffering and anguish as well as achieve economy of operations.

Just like in any other area of engineering, cost is also a very important factor in safety. In fact, this was basically due to engineering systems' cost; in the early years of Industrial Revolution, the factor of safety was generally overlooked. For example, once a railroad executive remarked: "It would cost less to bury a person killed in an accident than to put air brakes on a car" [6]. Nowadays, one of the major factors for the increasing attention on safety is the accidents' cost. This cost includes items such as lawsuit expenses, medical expenses, property damage amount, productivity losses, and wage losses.

This chapter presents various important aspects of safety management and costing.

5.2 SAFETY MANAGEMENT-RELATED PRINCIPLES AND DEVELOPING A SAFETY PROGRAM PLAN

There are many safety management-related principles. Some of the important ones are as follows [7, 8]:

- Management has the responsibility for making changes to the environment that leads to unsafe behavior because unsafe behavior is the result of normal people reacting to their surrounding environment.
- Circumstances that can be predicted to result in serious injuries are high energy sources, non-productive activities, certain construction situations, and non-routine activities.
- Management should manage safety just like managing any other function in the organization/company (i.e., by setting attainable goals, planning, organizing, and controlling).
- A key for having an effective safety system is clearly visible in the top management support, flexibility, worker participation, etc.

DOI: 10.1201/9781003365204-5

- The key for successful line safety performance is management procedures that factor in accountability in an effective manner.
- Symptoms that indicate something is wrong in the safety management system are an accident, an unsafe act, and an unsafe condition.
- In developing an effective safety system, carefully consider three major subsystems, that is, the physical, the behavioral, and the managerial.
- The function of safety is finding and defining accident-causing operational errors.
- Causes that lead to unsafe behavior can be highlighted, classified, and controlled.
- The safety system should be tailored to fit the company/organization culture.

The development of an appropriate safety plan is essential for organizations contemplating on introducing a safety program. Seven useful steps for developing such a plan are as follows [1]:

- **Step I: Develop and announce the safety policy.** This step is concerned with preparing and announcing the policy for controlling hazards within the organization as well as designating authority and accountability for its appropriate implementation.
- **Step II: Appoint safety chief.** This step is concerned with appointing a person with proper qualifications and experience for looking after all safety-associated matters within the organization.
- **Step III: Analyze operational injury records.** This step is concerned with conducting analysis of operational records of injuries, work-associated illnesses, and property damage.
- **Step IV: Evaluate operational hazards' scope and seriousness.** This step is concerned with items such as evaluating the quality of the existing physical safeguards, determining the nature and severity of inherent operating hazards, determining the corrective measures needed, and developing time estimates and budgets for conducting the corrective measures.
- **Step V: Select and schedule appropriate communication methods.** This step is concerned with selecting and scheduling appropriate communication approaches for various purposes including providing safety training to all involved workers and informing higher management about the organization's safety progress and related requirements.
- **Step VI: Establish a schedule for periodic reviews.** This step is concerned with developing a schedule for periodic reviews of the program and facilities in regard to safety.
- **Step VII:** This step is concerned with developing appropriate long-range and short-range objectives for the safety program.

5.3 SAFETY DEPARTMENT FUNCTIONS

A safety department conducts various types of functions, and these functions can vary from one organization to another. Nonetheless, 12 of the typical functions of a safety department are as follows [1, 9]:

- **Function I:** Evaluate company compliance with governmental regulations.
- **Function II:** Justify safety measures and publicize safety-associated materials.
- **Function III:** Develop the organization safety program.
- **Function IV:** Prepare reports on company safety program.
- **Function V:** Conduct safety inspections and surveys.
- **Function VI:** Manage the organization safety program.
- **Function VII:** Keep data on work injury and illness.
- **Function VIII:** Distribute personal protective equipment.
- **Function IX:** Investigate accidents.
- **Function X:** Provide safety training.
- **Function XI:** Procure personal protective equipment.
- **Function XII:** Liaise with others on safety matters.

5.4 QUALIFICATIONS AND FUNCTIONS OF SAFETY PROFESSIONALS

Two key professionals who play a very important role in an organization's safety program are safety engineer and safety manager. Qualifications and functions of these two professionals are presented below, separately.

5.4.1 SAFETY ENGINEER

A safety engineer must possess certain academic and other qualifications in order to conduct his/her job effectively. In 1992, the American Society of Safety Engineers (ASSE) conducted a survey of its 27,000 members with respect to their academic-related qualifications. The results of the survey are as follows [5, 7]:

- Doctorate degree (3%)
- Master's degree (25%)
- Bachelor's degree (75%)
- Majored in safety/health in college/university (30%)
- Obtained degrees in business/management (19%)
- Trained as engineers (20%)

Other qualifications include a good knowledge of safety matters, readiness for trying new ideas and approaches, easy to get along with safety related experience, enthusiasm, drive, and perseverance.

Some of the functions conducted by a safety engineer are as follows [9]:

- Perform safety studies.
- Perform safety inspections.
- Provide safety training.
- Collaborate with safety committees.
- Coordinate with management on matters related to safety.
- Ensure that the appropriate corrective measures are taken to avoid accident re-occurrences.

- Process workers' compensation claims.
- Conduct accident investigations.

5.4.2 SAFETY MANAGER

A safety manager must possess certain qualifications in order to conduct his/her task in an effective manner. Some of the qualifications are as follows [1, 5, 8, 10]:

- Appropriate academic qualifications.
- Good knowledge of business administration.
- Open mind to try new ideas and methods.
- Effective knowledge in safety and safety-related experience.
- Good supervisory skills.
- Effective knowledge in engineering principles.
- Ability to get others to do their tasks in an effective manner.
- Good drive, enthusiasm, and perseverance.

Some of the functions performed by a safety manager are as follows [1, 5, 8, 10]:

- Managing and formulating the safety program.
- Acquiring the latest hazard control information.
- Supervising department employees.
- Directing the inspection of the facilities for compliance with the regulations of the outside bodies.
- Directing the collection and recording of pertinent information on matters such as work injuries and accidents.
- Participating in procurement specification reviews.
- Representing management to public, government agencies, insurance companies, etc., with respect to safety.
- Participating in the design of new facilities/equipment layout/process layout.
- Reporting to higher management periodically with respect to the state of the company's safety effort.
- Advertising safety issues at all levels of management.

5.5 SAFETY COMMITTEES AND MOTIVATING EMPLOYEES TO WORK SAFELY

Safety committees play an instrumental role in promoting safety within an organization because they provide a systematic structure for funneling health and safety issues by workers and management. Prior to starting a safety committee, careful attention must be paid to items such as scope/responsibility of the committee, the committee's authority, and procedures.

The procedures are concerned with items such as the authority to whom committee reports are to be submitted, time and place of meetings, records to be retained, meeting frequency, and attendance requirements [1, 5, 11]. Past experiences over the years clearly indicate that success or failure of a safety committee largely depends on its composition. It means that for the committee's success, the members of committee must be from a broad cross-section representing all units within the organization. This will provide all employees a representative voice on the committee.

Nonetheless, the committee's health and safety professional member serves as a facilitator, advisor, and catalyst. The members of the committee appoint a chairperson and a secretary for recording the minutes. Past experiences over the years clearly indicate that normally best results are obtained, when neither the executive manager nor the health and safety professional serves as a committee chairperson. In addition, all members of the safety committee are trained in areas such as group problem-solving methods and cause-and-effect analysis [12].

Some of the important duties of a safety committee are as follows [5, 6]:

- Providing assistance in safety deficiency investigations.
- Providing assistance in safety training.
- Conducting inspection tours.
- Providing assistance in selecting safety-related posters.
- Providing assistance to safety professionals.
- Making safety recommendations.
- Promoting accident prevention.

Supervisors play an instrumental role in motivating employees to work safely. They can motivate employees directly or indirectly through actions as follows [5, 7]:

- Strengthening and enhancing management standards' importance for safe performance.
- Focusing attention on the importance of safe performance.
- Spending more time recognizing and rewarding safe performance than spending time in disciplining involved employees for unsafe performance.
- Reminding workers regularly about the methods of safe performance.

5.6 A MANUFACTURER'S LOSSES OR COST DUE TO ACCIDENT INVOLVING ITS PRODUCT

Losses or cost of a manufacturer due to an accident involving its product can occur in many different ways. Some of these ways are as follows [5, 13]:

- Loss of revenue because of degradation in public confidence.
- Loss of time of manufacturer's employees involved with the accident.
- Payments for death/injury claim settlements.
- Payments for property damage claims that are not covered by insurance.

- Cost of accident investigation.
- Cost of slowdown in operations while accident causes are being investigated.
- Cost of corrective measures needed to prevent accident reoccurrences.

5.7   SAFETY COST ESTIMATION MODELS AND METHODS

There are many models and methods used for estimating various types of safety costs. This section presents three such methods and models (i.e., one model and two methods).

5.7.1   TOTAL SAFETY COST ESTIMATION MODEL

In this model, the total safety cost is expressed by [5, 6]:

$$\text{TSC} = C_i + C_{il} + C_w + C_a + C_{im} + C_s + C_r + C_m \tag{5.1}$$

where

TSC is the total safety cost.
C_i is the cost of insurance.
C_{il} is the cost of immediate losses due to accidents.
C_w is the cost of welfare issues.
C_a is the cost of accident prevention measures.
C_{im} is the cost of immeasurables.
C_s is the cost of safety-associated legal issues.
C_r is the cost of rehabilitation and restoration.
C_m is the cost of miscellaneous safety issues.

5.7.2   THE HEINRICH METHOD

The method is named after H.W. Heinrich who categorically pointed out that for each and every insured cost dollar paid for accidents, there were approximately four dollars of uninsured cost borne by the company or organization [14]. His conclusions were based on factors listed below [5, 15].

- Examination of 5000 case files from organizations insured with a private company.
- Performance of research studies in those organizations.
- Interviews with these enterprises' administrative and production service staff.

Heinrich expressed "total cost of occupational injuries" as the sum of two elements: direct cost and indirect cost. The direct cost is the total benefits paid by the insurance company, whereas the expenditure assumed directly by the enterprise is the indirect cost. More specifically, the direct cost is composed of elements listed below [5, 15].

- Cost of injured worker's lost time.
- Cost of material/machine damage.
- Lost time cost of workers who stop their work and get involved in the action.
- Cost of lost orders.
- Cost of overheads for injured worker while in nonproduction mode.
- Cost of management's lost time.
- Cost associated with profit and worker productivity loss.
- Lost time cost of first aid and hospital workers not paid by insurance.
- Cost of weakened morale.

5.7.3 THE SIMONDS METHOD

The method was developed by Professor R.H. Simonds of Michigan State College by reasoning that an accident's cost is composed of two main elements, that is, insured cost and uninsured cost [5, 16]. Furthermore, he argued that the insured cost can easily be estimated by simply examining some accounting records or data, but the estimation of the uninsured cost is much more challenging or a trying task. Thus, for estimating the total uninsured cost of accidents, Simonds proposed the following equation [4, 16, 17]:

$$C_{tu} = \gamma_A C_A + \gamma_B C_B + \gamma_c C_c + \gamma_D C_D \tag{5.2}$$

where

C_{tu} is the total uninsured cost of accidents.
C_A is the average uninsured cost associated with Class A accidents.
γ_A is the total number of lost work-day cases due to Class A accidents resulting in permanent partial disabilities and temporary total disabilities.
C_B is the average uninsured cost associated with Class B accidents.
γ_B is the total number of physician's cases associated with Class B accidents. More specifically, the Occupational Safety and Health Act (OSHA) nonlost work day cases that require treatment by a doctor.
C_C is the average uninsured cost associated with Class C accidents.
γ_C is the total number of first-aid cases associated with Class C accidents. More specifically, those accidents in which first-aid was provided locally and resulting in a loss of less than one-eighth of working time.
C_D is the average uninsured cost associated with Class D accidents.
γ_D is the total number of noninjury cases associated with Class D accidents. More specifically, those accidents causing only minor injuries that do not need the attention of a medical professional.

5.8 SAFETY COST PERFORMANCE MEASUREMENT INDEXES

These indexes are used for measuring the overall safety cost performance of organizations. The main objective of using these indexes is to indicate trends, using the past safety cost performance as a reference point, and to encourage all involved individuals for improving over the past performance.

This section presents three most useful safety cost-associated indexes [11, 15, 17].

5.8.1 INDEX I

This index is concerned with determining the average cost per occupational injury in an organization and is defined by

$$AC_{oi} = \frac{C_{toi}}{OI_t} \tag{5.3}$$

where

AC_{oi} is the average cost per occupational injury.
C_{toi} is the total cost of occupational injuries expressed in dollars.
OI_t is the total number of occupational injuries.

5.8.2 INDEX II

This index is concerned with determining the average cost of occupational injuries per profit dollar in an organization and is defined by

$$AC_{pd} = \frac{C_{toi}}{TPD} \tag{5.4}$$

where

AC_{pd} is the average cost of occupational injuries per profit dollar.
TPD is the total profit expressed in dollars.

5.8.3 INDEX III

This index is concerned with determining the average cost of occupational injuries per unit turnover and is defined by

$$AC_{ut} = \frac{C_{toi}}{UT_t} \tag{5.5}$$

where

AC_{ut} is the average cost of occupational injuries per unit turnover.
UT_t is the total number of units turnover (i.e., unit quantity, tons, etc.).

PROBLEMS

1. Define the term "Safety management."
2. Discuss at least eight safety management-related principles.
3. List at least nine typical functions of a safety department.
4. Discuss the steps involved in developing a safety program plan.
5. What are the important functions performed by a safety engineer?
6. What are the important functions performed by a safety manager?
7. What are a manufacturer's losses or cost due to an accident involving its product?
8. What are the important duties of a safety committee?
9. Discuss the following two safety cost estimation methods:
   - The Heinrich method
   - Simonds method
10. Define two indexes used in measuring the overall performance of organizations with respect to injury cost.

REFERENCES

1. Grimaldi, J.V., Simonds, R.H., *Safety Management*, Richard D. Irwin, Chicago, 1989.
2. Petersen, D., *Safety Management: A Human Approach*, Aloray Publishers, Englewood Cliffs, NJ, 1975.
3. Roland, H.E., Moriarty, B., *System Safety Engineering and Management*, John Wiley and Sons, New York, 1983.
4. Goetsch, D.L., *Occupational Safety and Health*, Prentice Hall, Englewood Cliffs, NJ, 1996.
5. Dhillon, B.S., *Reliability, Quality, and Safety for Engineers*, CRC Press, Boca Raton, Florida, 2005.
6. Hammer, W., Price, D., *Occupational Safety Management and Engineering*, Prentice Hall, Upper Saddle River, NJ, 2001.
7. Petersen, D., *Safety Management*, American Society of Safety Engineers, Des Plaines, IL, 1998.
8. Dhillon, B.S., *Engineering Safety: Fundamentals, Techniques and Applications*, World Scientific Publishing, River Edge, NJ, 2003.
9. Gloss, D.S., Wardle, M.G., *Introduction to Safety Engineering*, John Wiley and Sons, New York, 1984.
10. Schenkelbach, L., *The Safety Management Primer*, Dow Jones-Irwin, Homewood, IL, 1975.
11. Blake, R.P., Ed., *Industry Safety*, Prentice Hall, Englewood Cliffs, NJ, 1964.
12. Peters, T., *Thriving on Chaos: Handbook for a Management Revolution*, Harper and Row, New York, 1987.
13. Hammer, W., *Product Safety Management and Engineering*, Prentice Hall, Englewood Cliffs, NJ, 1980.
14. Heinrich, H.W., *Industrial Accident Prevention*, McGraw Hill, New York, 1931.
15. Andreoni, D., *The Cost of Occupational Accidents and Diseases*, International Labour Office, Geneva, Switzerland, 1986.
16. Simonds, R.M., *Estimating Accident Cost in Industrial Plants*, Safety Practices Pamphlet No. III, National Safety Council, Chicago, 1950.
17. Raouf, R., Dhillon, B.S., *Safety Assessment: A Quantitative Approach*, Lewis Publishers, Boca Raton, FL, 1994.

6 Maintenance, Software, and Robot Safety

6.1 INTRODUCTION

Each year, a vast sum of money is being spent worldwide to design, develop, operate, and maintain various types of engineering systems, including robots and software. Their safety has become an important issue.

Accidents occurring during maintenance work or concerning maintenance are quite frequent. For example, in 1994, 13.6% of all accidents in the U.S. mining industry occurred during the maintenance activity and, since 1990, the occurrence of such accidents has increased each year. Therefore, it is essential that maintenance engineering should strive to eliminate or control potential safety hazards to ensure satisfactory protection to humans and material from things such as electrical shock, toxic gas sources, and moving mechanical assemblies [1, 2].

Nowadays, computers are used widely used in various types of applications, and they are composed of both hardware and software elements. In fact, nowadays, much more money is spent in developing computer software than in developing hardware, in comparison to the early computers. For example, in 1955, software accounted for approximately 20% of the total computer cost, and in the mid-1980s, the percentage increased to around 90% [3]. Needless to say, safety has become a very important issue in software because proper functioning of software is so important that a simple malfunction can result in a large-scale loss of human lives. For example, commuter trams in Paris, France, serve around 800,000 passengers daily and they very much depend on software signaling [4].

Nowadays, millions of robots are being used worldwide. The awareness of robot safety appears to have its beginning in the 1970s [5]. In the 1980s, three important documents concerning robot safety appeared: a technical guidance document [6], a standard [7], and a book (edited) [8]. In the 1990s, a book entitled *Robot Reliability and Safety* discussed robot safety in considerable depth [5]. A comprehensive list of publications on robot safety is given in Ref. [9].

This chapter presents various important aspects of maintenance, software, and robot safety.

6.2 MAINTENANCE SAFETY

Accidents occurring during the maintenance activity are quite frequent. This section presents four important aspects of maintenance safety.

DOI: 10.1201/9781003365204-6

6.2.1 FACTS, FIGURES, AND EXAMPLES

Some of the facts, figures, and examples concerned with maintenance safety are as follows:

- In 1994, 13.6% of the accidents in the U.S. mining industry occurred during maintenance [2].
- During the period 1978–1988, maintenance activity accounted for around 34% of all lost-time injuries in the U.S. mining industrial sector [10].
- In 1990, a steam leak occurred in the fire room on board the U.S.S. Iwo Jima (LPH2) naval ship, resulting in ten fatalities. Subsequently, an investigation revealed that a valve had just been repaired and bonnet fasteners were replaced with mismatched and wrong material [4].
- In 1991, an explosion at an oil refining company in Louisiana that killed four workers occurred as three gasoline synthesizing units were being brought back to their active state after some maintenance actions [12].
- In 1990, a newly replaced windscreen of a British BAC1-11 jet blew out, as the aircraft was climbing to its cruising altitude because of incorrect installation of the windscreen by a maintenance worker [13, 14].
- For the period 1982–1991, a study of safety issues with respect to onboard fatality of worldwide jet fleet revealed that maintenance and inspection was the second most important safety issue with a total of 1481 onboard fatalities [15, 16].
- In 1985, 520 people lost their lives in a Japan Airlines Boeing 747 jet accident due to an improper repair [14, 17].
- In 1979 in a DC-10 aircraft accident in Chicago, 272 persons lost their lives because of incorrect procedures followed by maintenance personnel [18].

6.2.2 REASONS FOR SAFETY-RELATED PROBLEMS IN MAINTENANCE AND FACTORS RESPONSIBLE FOR DUBIOUS SAFETY REPUTATION IN MAINTENANCE WORK

Past experiences over the years clearly indicate that there is a significant proportion of the total accidents that occur during the maintenance activity. Some of the important reasons for safety-related problems in the maintenance are as follows [2]:

- Poorly written maintenance instructions and procedures.
- Insufficient time for performing required maintenance task.
- Inadequate training to maintenance personnel.
- Inadequate safety tools and standards.
- Inadequate equipment design.
- Poor management.
- Inadequate work tools.
- Poor work environment.

There are many factors responsible for dubious safety reputation in maintenance work. Nine of these factors are as follows [19]:

- **Factor I:** Maintenance work conducted in unfamiliar surroundings means that hazards such as missing gratings, rusted handrails, and broken light fittings may pass unnoticed.
- **Factor II:** Need to transfer bulky and heavy materials from a warehouse to the maintenance workplace, sometimes using lifting and transport equipment well outside a strict maintenance regime.
- **Factor III:** Numerous maintenance tasks occur infrequently, for example, machinery failures, thus fewer opportunities to discern safety-associated problems and to introduce remedial measures.
- **Factor IV:** Performance of maintenance work underneath or inside machines such as large rotating machines, air ducts, and pressure vessels.
- **Factor V:** Some maintenance work may require conducting tasks such as disassembly of corroded parts, or manhandling cumbersome heavy parts in confined spaces and poorly lit areas.
- **Factor VI:** Performance of maintenance work at odd hours, in remote locations, and in small numbers.
- **Factor VII:** Disassembling previously working machinery, thus working under the risk of releasing stored energy.
- **Factor VIII:** Difficulty in maintaining regular communication with workers in some maintenance tasks.
- **Factor IX:** Sudden need for maintenance work, allowing limited time to prepare.

6.2.3 MAINTENANCE PERSONNEL SAFETY AND TYPICAL HUMAN BEHAVIORS

Generally, emphasis is placed on designing safety into machines rather than on the safety of the maintainers, operators, etc. On occasion, more protection is needed for maintenance personnel beyond the safety designed into machines or processes. Two important areas of maintenance worker safety are respiratory protection and protective clothing. Some of the important areas requiring respiratory protection are as follows [2]:

- Airborne contaminants, toxic gas, vapors, or dust
- Flames or radiation
- Air at extreme temperatures
- Oxygen-deficient air

The protective clothing includes items such as follows [19]:

- **Gloves:** These are important for protecting hands from injury when conducting various types of maintenance tasks.
- **Boots and toecaps:** Well-fitting boots with steel toecaps can reduce the risk of injury in situations such as dismantling used equipment where heavy metal parts are quite difficult to hold as well as are likely to slip and drop on exposed feet.

- **Hard hats and helmets:** These are useful for protecting maintenance workers from head injury.
- **Goggles, screens, safety glasses, and visors:** These items protect eyes from sparks, flying chips, jetted hydraulic fluid, chemical sprays, etc.
- **Ear defenders:** These are necessary where machine or process noise can damage maintenance workers' ears.

Over the years, professionals working in the maintenance area have observed various typical human behaviors that may impact safety in maintenance. Some of these behaviors are as follows [20–22]:

- Humans often overestimate the probability of occurrence of the "pleasant event" and underestimate the probability of occurrence of the "unpleasant event."
- Humans tend to use their hands first to test or examine.
- Humans tend to think of manufactured items as being safe.
- Humans frequently overestimate speed of an accelerating object and underestimate speed of a decelerating object.
- Humans normally know very little about their physical limitations.
- Humans generally continue to use faulty equipment/systems/items.
- Humans often overestimate bulky weight and underestimate compact weight.
- Humans are easily confused by unfamiliar items.

6.2.4 Useful Guidelines for Equipment Designers to Improve Safety in Maintenance

Over the years, professionals working in the area of maintenance have developed various guidelines for equipment designers to improve safety in maintenance. Some of these useful guidelines are as follows [2, 23]:

- Incorporate devices or other measures to permit early detection or prediction of potential failures so that maintenance can be conducted prior to actual failure with somewhat lower risk of hazard.
- Develop the design in such a way to lower the probability of maintenance workers being injured by electric shock, contact with a hot surface, escaping high-pressure, etc.
- Provide appropriate guards against moving parts and interlocks to block access to hazardous locations.
- Simplify the design as much as possible because complexity usually adds to maintenance problems.
- Eliminate the opportunity to perform maintenance or adjustments close to hazardous operating items.
- Develop designs or procedures for minimizing the occurrence of maintenance errors.
- Consider the typical human behaviors presented in Section 6.2.3.

- Provide fail-safe designs for preventing injury or damage in the event of a failure.
- Eliminate or minimize the need for special equipment/tools.

6.3 SOFTWARE SAFETY

Software safety may simply be described as the freedom from software-associated hazards. This section presents four important aspects of software safety.

6.3.1 FACTS AND FIGURES

Some of the facts and figures concerned with software safety are as follows:

- A software error caused a radioactive heavy water spill at a Canadian nuclear power generating station [24].
- A software error in a computer-controlled therapeutic radiation machine called Therac 25 resulted in deaths of two patients and severe injuries to another patient [25–27].
- A software error in a French meteorological satellite caused the destruction of 72 of the 141 weather balloons [28].
- A SAAB JAS39 Gripen fighter plane crashed due to an instrument failure caused by a safety-associated software issue [24].
- Over 70% of the companies involved in software development develop their software products by using ad hoc and unpredictable approaches [29].

6.3.2 SOFTWARE SAFETY VERSUS RELIABILITY

Quite often, software safety and reliability are equated, thus leading to a considerable degree of confusion. In order to avoid such confusion, their clear understanding is very important. In a broader perspective, it may simply be stated that the reliability-related requirements are concerned with making a system failure free and the safety mishap free [30, 31]. More specifically, reliability is concerned with all possible software errors and safety with only those errors that may result in system hazards.

Nonetheless, it may be added that not all software functioning as per specification is safe nor all soft-errors cause safety problems [32]. For example, in the past, severe mishaps have occurred while something (i.e., software) was operating exactly as per the stated requirement (i.e., without failure) [33, 34].

6.3.3 SOFTWARE HAZARD CAUSING WAYS

Software can cause or contribute to hazards in many different ways. Seven of these ways are as follows [31, 35]:

- **Way I:** By conducting a function that is not required.
- **Way II:** By failing to recognize a hazardous situation requiring a corrective measure.

- **Way III:** By conducting a function out-of-sequence.
- **Way IV:** By failing to conduct a required function.
- **Way V:** By poorly timing a response for an adverse situation.
- **Way VI:** By providing an incorrect solution to a given problem.
- **Way VII:** By responding poorly to a contingency.

6.3.4   Software Hazard Analysis Methods

There are many methods and techniques that can be used for performing various types of software hazard analysis [36, 37]. These include software sneak circuit analysis, proof of correctness, code walk-throughs, hazard and operability studies, software fault tree analysis, design walk-throughs, event-tree analysis, Petri net analysis, failure modes and effect analysis, structural analysis, cause-consequence diagrams, cross-reference-listing-analysis, and nuclear safety cross-check analysis.

The first three of these methods are described below [38].

6.3.4.1   Software Sneak Circuit Analysis

This method highlights software logic that causes undesired outputs. More specifically, program source code is converted to topological network trees, and patterns such as entry dome, return dome, single line, trap, iteration loop, and parallel line are utilized to model the code. Each and every software mode is modeled utilizing the basic patterns linked in a network tree flowing from top to bottom.

Four types of software sneaks are searched by the involved analysts: the undesired inhibit of an output, presence of an undesired output, a program message poorly describing the actual condition, and wrong timing. At the discovery of sneaks, analysts carry out investigative analyses to verify that the code under consideration does indeed produce the sneaks. At the end, the sneaks' impact is assessed and appropriate corrective measures are recommended. Additional information on the method is available in Refs. [36–38].

6.3.4.2   Proof of Correctness

This method decomposes a program under consideration into a number of logical segments and defines each segment input/output assertions. Subsequently, involved software professionals verify from the perspective that each and every input assertion and its associated output assertion are true and that if all of the input assertions are true, then all output assertions are also true.

Furthermore, this method makes use of mathematical theorem proving concepts to verify that a given program is consistent with its all associated specifications. Additional information on the method is available in Refs. [36–39].

6.3.4.3   Code Walk-Throughs

This is a quite effective method used for improving software safety and quality and basically is a team effort among various professionals: software programmers, software engineers, program managers, system safety specialists, etc.

Nonetheless, code walk-throughs are in-depth review of a given software through inspection and discussion regarding its functionality. More specifically, all logic branches and the function of each statement are discussed in a significant depth. Thus, the process provides a checks and balances system of the software produced. Additional information on the method is available in Ref. [39].

6.4 ROBOT SAFETY

Robot safety may simply be described as preventing the robot from damaging its environment, particularly the human element of that environment or simply preventing damage to the robot itself [40]. Four important aspects of robot safety are presented below.

6.4.1 FACTS AND FIGURES

Some of the directly or indirectly robot safety-associated facts and figures are as follows:

- A study reported that around 12–17% of the accidents in the industrial sector using advanced manufacturing technology were associated with automated production equipment [41, 42].
- A study of 32 robot-associated accidents in the United States, Japan, Germany, and Sweden reported that line workers were at the greatest risk of injury followed by maintenance personnel [43].
- During the period 1978–1987, a total of ten fatal robot accidents took place in Japan alone [44].
- During the period 1978–1984, a total of five fatal robot accidents occurred (i.e., four in Japan and one in the United States) [5].
- In 1978, the first robot-induced fatal accident took place in Japan [44].
- In 1984, the first fatal robot accident took place in the United States [45].

6.4.2 ROBOT SAFETY-RELATED PROBLEMS

Over the years, professionals working in the area of robotics have highlighted many safety-related problems unique to robots. Some of these problems are as follows [46]:

- Robots are prone to occurrence of safety-associated electrical design problems such as poorly designed power sources, potential electric shock, and fire hazards.
- The presence of a robot generally receives a great deal of attention from humans. These humans are quite often ignorant of the potential-associated hazards.
- A robot may go out of its programmed zones, strike something or throw objects in the event of a mechanical, hydraulic, or control failure.
- A robot may lead to a quite high risk of fire or explosion if it is installed in unsuitable environments.
- Mechanical design-related problems of robots may result in hazards such as pinching, grabbing, and pinning.

- Robot maintenance procedures can lead to various hazardous conditions.
- A robot creates potentially hazardous conditions because it often manipulates objects of different sizes and weights.

6.4.3   Types of Robot Accidents

Robot accidents may be grouped under the following four categories [47]:

- **Category I: Trapping/Crushing accidents.** These accidents are concerned with conditions where a person's limb or other body part is trapped between peripheral equipment and the robot arm or the person is physically driven into and crushed by other peripheral equipment.
- **Category II: Collision/Impact accidents.** These accidents are concerned with conditions where unpredicted movements, part failures, or unpredicted program changes related to the arm of robot/peripheral equipment result in contact accidents.
- **Category III: Mechanical part accidents.** These accidents include conditions where the breakdown of the robot's peripheral equipment, drive components, power source, or its tooling or end-effectors occurs. Two typical examples of mechanical failures are the failure of power tools or end-effectors and the failure of gripper mechanism.
- **Category IV: Miscellaneous accidents.** These accidents are all those accidents that cannot be grouped under the above three categories. Some examples of these accidents are environmental accidents from dust, arc flash, metal spatter, electromagnetic, or radio-frequency interference and tripping-related hazards from equipment and power cables on the floor.

6.4.4   Robot Safeguard Methods

There are many robot safeguard methods. Six commonly used such methods are as follows [5, 48]:

- **Method I: Infrared light arrays.** This method is used for protecting humans from potential dangers in the operating zone of a robot. The commonly used linear arrays of infrared sources are known as light curtains. Although this method has proven to be quite effective but time to time false triggering may take place due to factors such as heavy dust, smoke, or flashing lights because of misalignment of system elements.
- **Method II: Intelligent systems.** This method uses intelligent control systems for safeguarding. More clearly, these systems make use of avenues such as remote, sensing, software, and hardware in making decisions. Needless to say, an intelligent collision-avoidance system can only be achieved by restricting the robots' operating environment and using special software and sensors.
- **Method III: Flashing lights.** This method is concerned with installing flashing lights at the perimeter of the robot-working area or the robot itself.

Their purpose is to alert humans in the area that robot programmed motion is happening or could happen at any moment.

- **Method IV: Physical barriers.** This method uses items such as chain-link fences, plastic safety chains, and safety rails to stop humans from entering the restricted robot work zone.
- **Method V: Electronic devices.** This method is concerned with making use of active sensors for intrusion detection. Generally, the method is used in situations where the need of unobstructed floor space is important.
- **Method VI: Warning signs.** This method is normally used in circumstances where robots, by virtue of their size, speed, and inability to impart excessive force, cannot injure humans.

PROBLEMS

1. What are the important reasons for safety-related problems in maintenance?
2. List at least seven factors that are responsible for dubious safety reputation in maintenance work.
3. Discuss important items of protective clothing in maintenance work.
4. Discuss typical human behaviors with respect to safety in maintenance work.
5. Discuss at least seven useful guidelines for equipment designers to improve safety in maintenance.
6. What are the software hazard-causing ways? List at least seven of them.
7. List at least ten methods that can be used to perform software hazard analysis.
8. Describe the following two software hazard analysis methods:
   - Software sneak circuit analysis
   - Proof of correctness
9. Discuss at least seven robot safety-related problems.
10. Discuss at least five robot safeguard methods.

REFERENCES

1. AMCP 706-132, *Maintenance Engineering Techniques, U.S. Army Material Command*, Department of the Army, Washington, DC, 1975.
2. Dhillon, B.S., *Engineering Maintenance: A Modern Approach*, CRC Press, Boca Raton, Florida, 2002.
3. Keene, S.J., Software Reliability Concepts, *Annual Reliability and Maintainability Symposium Tutorial Notes*, 1992, pp. 1–21.
4. Cha, S.S., Management Aspect of Software Safety, *Proceedings of the Eighth Annual Conference on Computer Assurance*, 1993, pp. 35–40.
5. Dhillon, B.S., *Robot Reliability and Safety*, Springer-Verlag, New York, 1991.
6. An Interpretation of the Technical Guidance on Safety Standards in the Use, etc., of Industrial Robots, Japanese Industrial Safety and Health Association, Tokyo, 1985.
7. American National Standard for Industrial Robots and Robot Systems: Safety Requirements, ANSI/RIA R15.06-1986, American National Standards Institute (ANSI), New York, 1986.
8. Bonney, M.C., Yong, Y.F., Ed., *Robot Safety*, Springer-Verlag, New York, 1985.

9. Dhillon, B.S., *Robot System Reliability and Safety: A Modern Approach*, CRC Press, Boca Raton, Florida, 2015.

10. MSHA Data for 1978-1988, Mine Safety and Health Administration (MSHA), US Department of Labor, Washington, DC.

11. Joint Fleet Maintenance Manual, Vol. 5, Quality Maintenance, Submarine Maintenance Engineering, U.S. Navy, Portsmouth, NH.

12. Goetsch, D.L., *Occupational Safety and Health*, Prentice-Hall, Englewood Cliffs, New Jersey, 1996.

13. Report on the Accident to BAC 1-11, Report No. 1-92, Air Accident Investigation Branch, Ministry of Transportation, London, UK, 1992.

14. ATSB Survey of Licensed Aircraft Maintenance Engineers in Australia, Report No. ISBN 0642274738, *Australian Transport Safety Bureau (ATSB)*, Department of Transport and Regional Services, Canberra, Australia, 2001.

15. Human Factors in Airline Maintenance: A Study of Incident Reports, *Bureau of Air Safety Investigation*, Department of Transport and Regional Development, Canberra, Australia, 1997.

16. Russell, P.D., Management Strategies for Accident Prevention, *Air Asia*, Vol. 6, 1994, pp. 31–41.

17. Gero, D., *Aviation Disasters*, Patrick Stephens, Sparkford, UK, 1993.

18. Christensen, J.M., Howard, J.M., Field Experience in Maintenance, in *Human Detection and Diagnosis of System Failures*, Rasmussen, J., Rouse, W.B., Ed., Plenum Press, New York, 1981, pp. 111–133.

19. Stoneham, D., *The Maintenance Management and Technology Handbook*, Elsevier Science, Oxford, UK, 1998.

20. Nertnery, R.J., Bullock, M.G., Human Factors in Design, Report No. ERDA-76-45-2, The Energy Research and Development Administration (ERDA), Department of Energy, Washington, DC, 1976.

21. Woodson, W.E., *Human Factors Design Handbook*, McGraw-Hill, New York, 1981.

22. Dhillon, B.S., *Advanced Design Concepts for Engineers*, Technomic Publishing Co., Lancaster, Pennsylvania, 1998.

23. Hammer, W., *Product Safety Management and Engineering*, Prentice-Hall, Englewood Cliffs, New Jersey, 1980.

24. Mendis, K.S., Software Safety and Its Relation to Software Quality Assurance, in *Handbook of Software Quality Assurance*, G.G. Schulmeyer and J.E. McManus, Ed., Prentice Hall, Upper Saddle River, New Jersey, 1999, pp. 669–679.

25. Gowen, L.D., Yap, M.Y., Traditional Software Development's Effects on Safety, *Proceedings of the 6th Annual IEEE Symposium on Computer-Based Medical Systems*, 1993, pp. 58–63.

26. Joyce, E., Software Bugs: A Matter of Life and Liability, *Datamation*, Vol. 33, No. 10, 1987, pp. 88–92.

27. Schneider, P., Hines, M.L.A., Classification of Medical Software, *Proceedings of the IEEE Symposium Applied Computing*, 1990, pp. 20–27.

28. Anonymous, Blow Balloons, Aviation Week Space Technology, September 20, 1971, p. 17.

29. Thayer, R.H., Software Engineering Project Management, in *Software Engineering*, M. Dorfman and R.H. Thayer, IEEE Computer, Eds., Society Press, Los Alamitos, CA, 1997, pp. 358–371.

30. Ericson, C.A., Software and System Safety, *Proceedings of the 5th International System Safety Conference*, 1981, pp. III B.1–III B.11.

31. Levenson, N.G., Software Safety: Why, What, How, *Computing Surveys*, Vol. 18, No. 2, 1986, pp. 125–163.
32. Dhillon, B.S., *Engineering Safety: Fundamentals, Techniques, and Applications*, World Scientific Publishing, River Edge, New Jersey, 2003.
33. Leveson, N.G., Software Safety in Computer-Controlled Systems, IEEE Computer, February 1984, pp. 48–55.
34. Royland, H.E., Moriarty, B., *System Safety Engineering and Management*, John Wiley and Sons, New York, 1983.
35. Friedman, M.A., Voas, J.M., *Software Assessment*, John Wiley and Sons, New York, 1995.
36. Hammer, W., Price, D., *Occupational Safety Management and Engineering*, Prentice Hall, Upper Saddle River, New Jersey, 2001.
37. Ippolito, L.M., Wallace, D.R., A Study on Hazard Analysis in High Integrity Software Standards and Guidelines, Report No. NISTIR 5589, National Institute of Standards and Technology, U.S. Department of Commerce, Washington, D.C., January 1995.
38. Hensen, M.D., Survey of Available Software-Safety Analysis Techniques, *Proceedings of the Annual Reliability and Maintainability Symposium*, 1989, pp. 46–49.
39. Sheriffs, Y.S., Software Safety Analysis: The Characteristics of Efficient Technical Walkthroughs, *Microelectrics and Reliability*, Vol. 32, No. 3, 1992, pp. 407–414.
40. Graham, J.H., Overview of Robot Safety, Reliability, and Human Factors Issues, in *Safety, Reliability, and Human Factors in Robotic Systems*, J.H. Graham, Ed., Van Nostrand Reinhold, New York, 1991, pp. 1–10.
41. Backtrom, T., Doos, M., A Comparative Study of Occupational Accidents in Industries with Advanced Manufacturing Technology, *International Journal of Human Factors in Manufacturing*, Vol. 5, 1995, pp. 267–282.
42. Clark, D.R., Lehto, M.R., Reliability, Maintenance, and Safety of Robots, in *Handbook of Industrial Robotics*, S.Y. Nof, Ed., John Wiley and Sons, New York, 1999, pp. 717–753.
43. Jiang, B.C., Gainer, C.A., A Cause and Effect Analysis of Robot Accidents, *Journal of Occupational Accidents*, Vol. 9, 1987, pp. 27–45.
44. Nagamachi, M., Ten Fatal Accidents Due to Robots in Japan, in *Ergonomics of Hybrid Automated Systems*, W. Karwowski, et al., Ed., Elsevier, Amsterdam, 1988, pp. 391–396.
45. Sanderson, L.M., Collins, J.N., McGlothlin, J.D., Robot-Related Fatality Involving an U.S. Manufacturing Plant Employee: Case Report and Recommendations, *Journal of Occupational Accidents*, Vol. 8, 1986, pp. 13–23.
46. Van Deest, R., Robot Safety: A Potential Crisis, *Professional Safety*, January 1984, pp. 40–42.
47. Industrial Robots and Robot System Safety, Chapter 4, in *OSHA Technical Manual, Occupational Safety and Health Administration (OSHA)*, Department of Labor, Washington, DC, 2001.
48. Addison, J.H., Robotic Safety Systems and Methods: Savannah River Site, Report No. DPST-84-907 (DE-35-008261), December 1984, issued by E.I. du Pont de Nemours and Co., Savannah River Laboratory, Aiken, South Carolina.

7 Introduction to Maintainability

7.1 NEED FOR MAINTAINABILITY

Nowadays, the need for maintainability is becoming more important than ever before because of the alarming high operating and support costs of systems and equipment. For example, each year, the U.S. industrial sector spends over $300 billion on plant maintenance and operation [1]. Furthermore, the annual cost of maintaining a military jet aircraft is about $1.6 million; around 11% of the total operating cost for an aircraft is spent on maintenance-related activities [2].

Some of the objectives of applying maintainability principles are to reduce projected maintenance time and costs, to determine labor hours and other related sources needed for conducting the projected maintenance, and to use maintainability data for estimating equipment availability or unavailability.

When maintainability principles are applied successfully to any system/product, results such as reduction in system/product downtime, efficient restoration of the system/product to its operating state, and maximum operational readiness of the product/system can be expected [3, 4].

7.2 MAINTAINABILITY VERSUS MAINTENANCE

As maintainability and maintenance are closely interrelated, many people find it quite difficult to make a clear distinction between them. Maintainability refers to measures or steps taken during the system/product design phase for including features that will increase ease of maintenance and ensure that the system/product will have minimum downtime and lifecycle support costs when used in field environments [3]. In contrast, maintenance refers to the measures taken by the system/product users for keeping it in an operational state or repairing it to operational state [1, 5].

More simply, maintainability is a design parameter intended for minimizing system/equipment repair time, whereas maintenance is the act of servicing and repairing system/equipment [6].

The maintenance engineers' responsibility is to ensure that product or equipment design and development-related requirements reflect the maintenance needs of users. Thus, they are concerned with factors such as the environment in which the system/product will be operated and maintained; product and system mission, operational, and support profiles; and the levels and types of maintenance needed. Product/system maintainability design requirements are determined by various processes, including the analysis of maintenance tasks and requirements, the determination of maintenance resource requirements, the development of maintenance-related concepts, and maintenance engineering analysis [4].

DOI: 10.1201/9781003365204-7

7.3 MAINTAINABILITY'S SPECIFIC GENERAL PRINCIPLES

There are many maintainability's specific general principles. Some of the important ones are as follows [7]:

- Consider advantages of modular replacement versus part repair or throw-away design.
- Determine the extent of preventive maintenance to be carried out.
- Reduce the frequency, amount, and complexity of all required maintenance tasks.
- Eliminate or reduce the need for maintenance.
- Reduce the amount of supply supports needed.
- Reduce life cycle maintenance costs.
- Provide for maximum interchange ability.
- Reduce mean time to repair (MTTR).

7.4 MAINTAINABILITY FUNCTIONS

Just like in other areas of engineering, probability distributions also play an important role in maintainability. They are used for representing repair times of systems, equipment, and parts. After identification of the repair distribution, the corresponding maintainability function may be obtained. The maintainability function is concerned with predicting the probability that a repair, starting at time $t = 0$, will be completed in a time t.

Mathematically, the maintainability function is expressed by [3]

$$M(t) = \int_0^t f_r(t) \, dt \qquad (7.1)$$

where

$M(t)$ is the maintainability function.
t is time.
$f_r(t)$ is the probability density function of the repair time.

Maintainability functions for five probability distributions are obtained below [3, 4, 8–10].

7.4.1 MAINTAINABILITY FUNCTION FOR EXPONENTIAL DISTRIBUTION

This is a quite useful probability distribution to represent repair times and is simple and straightforward to handle. Its probability density function with respect to repair times is expressed by

$$f_r(t) = \theta e^{-\theta t} \qquad (7.2)$$

where

$f_r(t)$ is the repair time probability density function.
θ is the constant repair rate or reciprocal of the MTTR.
t is the variable repair time.

By substituting Equation (7.2) into Equation (7.1), we obtain

$$M_e(t) = \int_0^t \theta e^{-\theta t}\, dt = 1 - e^{-\theta t} \tag{7.3}$$

where

$M_e(t)$ is the maintainability function for exponential distribution.

Since $\theta = \dfrac{1}{MTTR}$, Equation (7.3) becomes

$$M_e(t) = 1 - e^{-\left(\frac{1}{MTTR}\right)t} \tag{7.4}$$

Example 7.1

Assume that the repair times to an engineering system are exponentially distributed with a mean value of 4 hours. Calculate the probability of completing a repair in 5 hours.
By substituting the given data values into Equation (7.4), we obtain

$$M(5) = 1 - e^{-\left(\frac{5}{4}\right)}$$
$$= 0.7134$$

This means that there is a likelihood of approximately 71% that the repair will be completed within 5 hours.

7.4.2 Maintainability Function for Rayleigh Distribution

This distribution is quite often used in reliability-related studies, and it can also be used to represent corrective maintenance times. Its probability density function with respect to corrective maintenance times (i.e., repair times) is expressed by

$$f_r(t) = \frac{2}{\gamma^2} t e^{-\left(\frac{t}{\gamma}\right)^2} \tag{7.5}$$

where

$f_r(t)$ is the repair time probability density function.
t is the variable repair time.
γ is the distribution scale parameter.

By inserting Equation (7.5) into Equation (7.1), we obtain

$$M_r(t) = \int_0^t \frac{2}{\gamma^2} t e^{-\left(\frac{t}{\gamma}\right)^2} dt$$

$$= 1 - e^{-\left(\frac{t}{\gamma}\right)^2}$$

(7.6)

where

$M_r(t)$ is the maintainability function for Rayleigh distribution.

7.4.3 MAINTAINABILITY FUNCTION FOR WEIBULL DISTRIBUTION

Sometimes, this distribution is used for representing corrective maintenance times, particularly for electronic equipment. Its probability density function with respect to corrective maintenance times is defined by

$$f_r(t) = \frac{\alpha}{\gamma^\alpha} t^{\alpha-1} e^{-\left(\frac{t}{\gamma}\right)^\alpha}$$

(7.7)

where

$f_r(t)$ is the corrective maintenance or repair time probability density function.
t is the variable repair time.
α is the distribution shape parameter.
γ is the distribution scale parameter.

By substituting Equation (7.7) into Equation (7.1), we obtain

$$M_w(t) = \int_0^t \frac{\alpha}{\gamma^\alpha} t^{\alpha-1} e^{-\left(\frac{t}{\gamma}\right)^\alpha} dt$$

$$= 1 - e^{-\left(\frac{t}{\gamma}\right)^\alpha}$$

(7.8)

where

$M_w(t)$ is the maintainability function for Weibull distribution.

At $\alpha = 1$ and $\alpha = 2$, Equation (7.8) reduces to Equation (7.3), for $\theta = \dfrac{1}{\gamma}$, and Equation (7.6), respectively.

7.4.4 MAINTAINABILITY FUNCTION FOR NORMAL DISTRIBUTION

This distribution is one of the most well-known probability distributions and it can also be used to represent failed equipment repair times. Its probability density function with respect to repair times is defined by

$$f_r(t) = \frac{1}{\sigma\sqrt{2\Pi}} \exp\left[-\frac{1}{2}\left(\frac{t-\theta}{\sigma}\right)^2\right]$$ (7.9)

where

$f_r(t)$ is the repair time probability density function.
t is the variable repair time.
θ is the mean of repair times.
σ is the standard deviation of the variable repair time t around the mean value of θ.

By substituting Equation (7.9) into Equation (7.1), we obtain

$$M_n(t) = \frac{1}{\sigma\sqrt{2\Pi}} \int_0^t \exp\left[-\frac{1}{2}\left(\frac{t-\theta}{\sigma}\right)^2\right] dt$$ (7.10)

where

$M_n(t)$ is the maintainability function for normal distribution.

The mean of repair times is given by

$$\theta = \sum_{i=1}^{m} t_i / m$$ (7.11)

where

m is the number of repair times.
t_i is the repair time i, for $i = 1, 2, 3, ..., m$.

The standard deviation is given by

$$\sigma = \left[\sum_{i=1}^{m}(t_i - \theta)^2 / (m-1)\right]^{1/2}$$ (7.12)

7.4.5 MAINTAINABILITY FUNCTION FOR GAMMA DISTRIBUTION

This distribution is one of the most flexible distributions and it can be used for representing various types of maintenance time data. Its probability density function with respect to repair times is expressed by

$$f_r(t) = \frac{b^n}{\Gamma(n)} t^{n-1} e^{-bt} \tag{7.13}$$

where

$f_r(t)$ is the repair time probability density function.
t is the variable repair time.
b is the distribution scale parameter.
n is the distribution shape parameter.

The gamma function, $\Gamma(n)$, is given by [3]

$$\Gamma(n) = \int_0^\infty y^{n-1} e^{-y} dy \tag{7.14}$$

By inserting Equation (7.13) into Equation (7.1), we obtain

$$M_g(t) = \frac{b^n}{\Gamma(n)} \int_0^t t^{n-1} e^{-bt} dt \tag{7.15}$$

where

$M_g(t)$ is the maintainability function for gamma distribution.

For $n = 1$, Equation (7.15) becomes the maintainability function for the exponential distribution. In order to find, $M_g(t)$, by using the tables of the incomplete gamma function, we rewrite Equation (7.15) to the following form [11]:

$$M_g(t) = b^n L(t) \tag{7.16}$$

where

$$L(t) = \frac{1}{\Gamma(n)} \int_0^t t^{n-1} e^{-bt} dt \tag{7.17}$$

The mean of the gamma distributed repair times is given by

$$\beta = \frac{n}{b} \tag{7.18}$$

where

β is the mean value of the gamma distributed repair times.

The standard deviation of the gamma distributed repair times is given by

$$\sigma = \frac{\sqrt{n}}{b} = \left(\frac{\beta}{b}\right)^{1/2} \tag{7.19}$$

where

σ is the standard deviation of the gamma distributed repair times.

PROBLEMS

1. Compare maintainability with maintenance.
2. Describe the need for maintainability.
3. List at least seven specific general principles of maintainability.
4. Define maintainability function.
5. Write down the maintainability function for an exponential distribution.
6. Assume that the repair times of an engineering system are exponentially distributed with a mean value of 3 hours. Calculate the probability of completing a repair in 4 hours.
7. Prove that the maintainability function for Rayleigh distribution is given by Equation (7.6).
8. Prove that the maintainability function for Weibull distribution is given by Equation (7.8).
9. What are the special case maintainability functions of the Weibull distribution?
10. Prove that the mean of the gamma distributed repair times is given by Equation (7.18).

REFERENCES

1. Latino, C.J., Hidden Treasure: Eliminating Chronic Failures Can Cut Maintenance Cost upto 60%, Report, Reliability Center, Hopewell, VA, 1999.
2. Kumar, U.D., New Trends in Aircraft Reliability and Maintenance Measures, *Journal of Quality in Maintenance Engineering*, Vol. 5, No. 4, 1999, pp. 287–295.
3. AMCP 706-133, *Engineering Design Handbook: Maintainability Engineering Theory and Practice*, Department of Defense, Washington, D.C., 1976.

4. Dhillon, B.S., *Engineering Maintainability*, Gulf Publishing, Houston, TX, 1999.
5. Downs, W.R., Maintainability Analysis Versus Maintenance Analysis, *Proceedings of the Annual Reliability and Maintainability Symposium*, 1976, pp. 476–481.
6. Smith, D.J., Babb, A.H., *Maintainability Engineering*, John Wiley and Sons, New York, 1973.
7. AMCP-706-134, *Engineering Design Handbook: Maintainability Guide for Design*, Department of Defense, Washington, DC, 1972.
8. Blanchard, B.S., Verma, D., Peterson, E.L., *Maintainability*, John Wiley and Sons, New York, 1995.
9. Von Alven, W.H., Ed., *Reliability Engineering*, Prentice Hall, Englewood Cliffs, New Jersey, 1964.
10. Dhillon, B.S., *Reliability Engineering in Systems Design and Operation*, Van Nostrand Reinhold, New York, 1983.
11. Pearson, K., *Tables of the Incomplete Gamma Function*, Cambridge University Press, Cambridge, UK, 1934.

8 Maintainability Tools and Specific Maintainability Design-Related Considerations

8.1 INTRODUCTION

Over the years, many methods and techniques have been developed for performing various types of reliability and quality analyses. Some of these methods and techniques have been successfully used in the area of maintainability as well. These methods and techniques include failure modes, effects, and criticality analysis; cause and effect diagram; total quality management (TQM); and fault tree analysis (FTA).

Past experiences over the years clearly indicate that an effective engineering design (i.e., cost-effective and supportable design) takes into account the maintainability considerations that arise during the equipment or item life cycle phases. This requires careful planning and a systematic effort for bringing attention to maintainability-related design factors such as maintainability design characteristics, maintainability allocation, and maintainability evaluation. Many of these factors involve subfactors such as standardization, interchangeability, accessibility, human factors, testing and checkout, safety, and modularization. In every aspect of maintainability design interchangeability, modularization, accessibility, and standardization are important considerations [1, 2].

This chapter presents a number of methods for conducting various types of maintainability analysis and various aspects of specific maintainability design considerations.

8.2 CAUSE-AND-EFFECT DIAGRAM

This is a deductive analysis method, and it can be useful in maintainability work. It is to be noted that in the published literature, the method is also known as a fishbone diagram because it resembles the skeleton of a fish, or as an Ishikawa diagram, after its founder, K. Ishikawa of Japan [3]. A cause-and-effect diagram uses a graphic fishbone for depicting the cause-and-effect relationships between an undesired event and its related contributing causes.

The diagram's right side represents the effect (i.e., the problem or the undesired event), and the left side represents that all possible causes of the problem are

DOI: 10.1201/9781003365204-8

connected to the central fish spine. The basic steps involved in developing a cause-and-effect diagram are as follows:

- **Step I:** Establish a problem statement or highlight the effect to be investigated.
- **Step II:** Brainstorm to highlight all possible causes for the problem under study.
- **Step III:** Group major causes into categories and stratify them.
- **Step IV:** Construct the diagram by linking the causes under appropriate process steps and write down the effect or problem in the diagram box (i.e., the fish head) on the right side.
- **Step V:** Refine all cause categories by asking questions such as "What causes this?" and "What is the reason for the existence of this condition?".

Some of the important advantages of the cause-and-effect diagram are as follows:

- Useful for presenting an orderly arrangement of theories.
- Useful to highlight (problem) root causes.
- Useful for generating ideas.
- Useful in guiding further inquiry.

A well-developed cause-and-effect diagram can be an effective tool to highlight possible maintainability-associated problems [1].

8.3 FAILURE MODES, EFFECTS, AND CRITICALITY ANALYSIS (FMECA)

This method grew out of failure mode and effects analysis (FMEA), described in Ref. [4]. When FMEA evaluates the failure criticality (i.e., the failure effect severity and its occurrence probability), the method is referred to as FMECA and the failure modes are assigned priorities [5].

Nine basic steps used for performing FMECA are as follows [5]:

- **Step I:** Understand system mission/operation/parts.
- **Step II:** Highlight the hierarchical or indenture level at which analysis is to be conducted.
- **Step III:** Highlight each item to be analyzed (e.g., subsystem, module, or part).
- **Step IV:** Establish appropriate ground rules and assumptions.
- **Step V:** Highlight all possible modes for each item under consideration.
- **Step VI:** Determine the effect of failure of each item for each failure mode.
- **Step VII:** Determine the effect of group failures on system mission and operation.
- **Step VIII:** Highlight methods/procedures for detecting potential failures.

- **Step IX:** Determine any design-associated changes/provisions that would prevent the occurrence of failures or mitigate their effects.

Some of the specific FMECA-related factors and their corresponding data retrieval sources in parentheses are as follows [1, 5]:

- Item nomenclature and functional specifications (design engineer or parts list).
- Item failure modes, causes, and rates (factory database or field experience database).
- Failure detection method (s) (design engineer or maintainability engineer).
- Item identification numbers (parts list for the product or system).
- Failure probability and severity classification (safety engineer).
- Mission phase and operational mode (design engineer).
- Failure effects (design engineer, safety engineer, or reliability engineer).
- Product and system function (customer requirements or the design engineer).
- Provisions and design changes to prevent or compensate for failures (design engineer).

Some of the advantages of FMECA are as follows [6]:

- A useful tool to highlight all possible failure modes and their effects on the mission, the system, and personnel.
- Useful for generating data for application in system safety and maintainability analyses.
- A quite effective tool for improving communication among design interface personnel.
- Useful for analyzing small, large, and complex systems effectively.
- Useful for generating input data for application in test planning.
- A systematic method for classifying hardware failures.
- Starts from the level of greatest detail and works in the upward direction.
- Quite useful for making design comparisons.
- Easy to understand.
- A visibility tool for managers and others.

8.4   FAULT TREE ANALYSIS (FTA)

This is described in detail in Chapter 4, and it can also be used to perform maintainability analysis [4, 7, 8]. Its application in maintainability work is demonstrated through the following two examples:

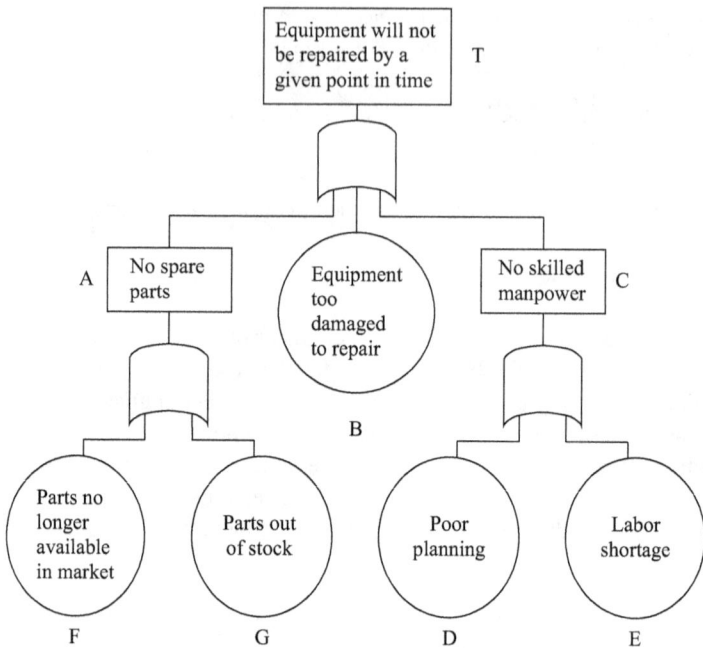

FIGURE 8.1   Fault tree for Example 8.1.

Example 8.1

Assume that a workshop that repairs failed engineering equipment, it will not be able to repair a given piece of equipment due to the following factors:

- **A:** There are no spare parts.
- **B:** Equipment is too damaged to repair.
- **C:** Skilled manpower is unavailable.

Furthermore, the unavailability of skilled manpower can be caused by either of the following two factors:

- **D:** Poor planning.
- **E:** Labor shortage.

In addition, either of the following two factors can result in the unavailability of spare parts:

- **F:** Parts are no longer available in the market.
- **G:** Parts are out of stock.

Develop a fault tree for this undesired event: the equipment will not be repaired by a given point in time.

For this example, the fault tree shown in Figure 8.1 was developed by using the symbols from Figure 4.1 presented in Chapter 4.

Example 8.2

Calculate the probability of the occurrence of the top event T (i.e., equipment will not be repaired by a given point in time) shown in Figure 8.1 if the probabilities of occurrence of factors B, D, E, F, and G are 0.01, 0.02, 0.03, 0.04, and 0.05, respectively. It is to be noted that the single capital letters in the Figure 8.1 diagram denote corresponding fault events.

By substituting the specified data values into Equation (4.1), the probability of occurrence of event A (i.e., no spare parts) is given by

$$P(A) = 1 - (1-F)(1-G) = 1 - (1-0.04)(1-0.05) = 0.088$$

Similarly, by substituting the given data values into Equation (4.1), the probability of occurrence of event C (i.e., skilled manpower is unavailable) is given by

$$P(C) = 1 - (1-D)(1-E) = 1 - (1-0.02)(1-0.03) = 0.0494$$

Finally, by substituting the above two calculated values and the given data value into Equation (4.1), the probability of the occurrence of the top event T (i.e., equipment will not be repaired by a given point in time) is given by

$$P(T) = 1 - (1-A)(1-B)(1-C) = 1 - (1-0.088)(1-0.01)(1-0.0494) = 0.1417$$

where

P(T) is the probability that the equipment will not be repaired by a given point in time.

Figure 8.2 presents the Figure 8.1 fault tree with given and calculated fault event occurrence probability values. As shown in Figure 8.2, the probability that the equipment will not be repaired by a given point in time is 0.1417.

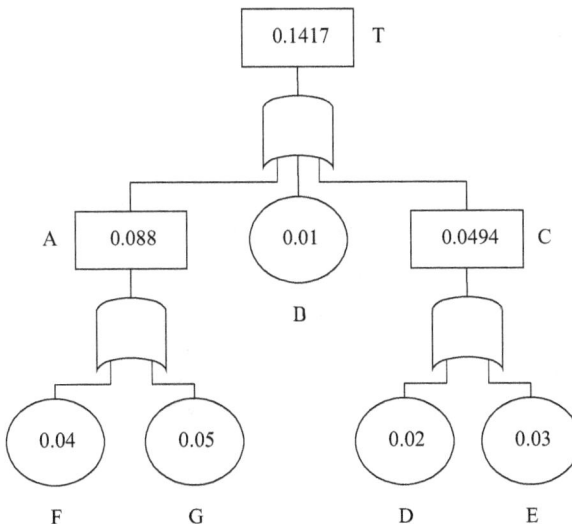

FIGURE 8.2   A fault tree for Example 8.1 with given and calculated fault event occurrence probability values.

8.5 TOTAL QUALITY MANAGEMENT (TQM)

TQM is a philosophy of pursuing continuous improvement in every process through the integrated or team efforts of all individuals in an organization. It has proven to be very useful to organizations in pursuit of improving the maintainability of their products. The term "total quality management" was coined in 1985 by an American behavioral scientist named Nancy Warren [9].

Two fundamental principles of TQM are customer satisfaction and continuous improvement, and its seven important elements are as follows [10]:

- Supplier participation
- Customer service
- Management commitment and leadership
- Statistical tools
- Cost of quality
- Team effort
- Training

TQM can be implemented by following the five basic steps presented below [1]:

- **Step I:** Create a vision.
- **Step II:** Plan an action.
- **Step III:** Create a structure (e.g., create cross-functional teams, eliminate roadblocks, institute appropriate training, and involve employees).
- **Step IV:** Measure progress.
- **Step V:** Update plans and visions as appropriate.

Many organizations have experienced various difficulties in implementing TQM. Some of these difficulties are failure of senior management to delegate decision-making authority to lower organizational levels, management insisting on implementing processes in a way employees find unacceptable, failure of top management to devote adequate time to the effort, and insufficient allocation of resources for training and developing manpower [11].

8.6 MAINTAINABILITY DESIGN FACTORS

There are many goals of maintainability design. They include increasing ease of maintenance, minimizing the logistical burden through resources (e.g., repair staff, support equipment, and spare parts) required for maintenance and support, reducing support costs, and minimizing preventive and corrective maintenance tasks [2].

Maintainability design factors that are most frequently addressed (ranked in descending order) are as follows [2]:

- Accessibility
- Test points
- Controls

- Labeling and coding
- Displays
- Manuals, checklists, charts, and aids
- Test equipment
- Tools
- Connectors
- Cases, covers, and doors
- Mounting and fasteners
- Handles
- Safety factors

The additional factors are standardization, modular design, interchangeability, ease of removal and replacement, lubrication, servicing equipment, skill requirements, indication and location of failures, work environment, required number of personnel, adjustments and calibrations, functional packaging, weight, cabling and wiring, fuses and circuit breakers, installation, illumination, training requirements, and test adopters and test hook-ups.

8.7 SIMPLIFICATION AND ACCESSIBILITY

Simplification is maintainability's most important element, and it is also probably the most difficult to achieve. Simplification should be design's constant objective, and a good design engineer includes all pertinent functions of a system/product into the design itself and makes use of as few components as good design-related practices will allow.

Accessibility is the relative ease with which an item can be reached for replacement, service, or repair. Poor accessibility is an often cause of ineffective maintenance. For example, as per a U.S. Army document, gaining access to equipment is probably second only to fault isolation as a time-consuming maintenance task [2].

There are many factors that affect accessibility. Some of these factors are as follows [2, 12]:

- The types of maintenance tasks to be performed through the access opening.
- Specified time requirements for performing the tasks.
- The visual needs of personnel performing the tasks.
- Work clearances necessary for conducting the tasks.
- The frequency with which the access opening is entered.
- The types of tools and accessories needed to perform the tasks.
- The item's location and environment.
- The clothing worn by the maintenance personnel.
- The distance to be reached to access the item.
- The danger associated with use of the access opening.
- The mounting of items behind the access opening.

Finally, it should be noted that an item being readily accessible does not in itself guarantee overall cost-effectiveness and ease of maintenance under consideration.

8.8 INTERCHANGEABILITY AND IDENTIFICATION

Interchangeability means including as an intentional aspect of design that any item can be replaced within a system/product by any similar item. Interchangeability is made possible through standardization and is a very important maintainability design factor. Three basic principles of interchangeability are as follows:

- **Principle I:** Strict interchangeability could be uneconomical in products/items that are expected to operate without any part replacement.
- **Principle II:** In products/items requiring frequent replacement and servicing of parts, each part must be interchangeable with another similar part.
- **Principle III:** Liberal tolerances must exist.

In order to achieve maximum interchangeability of parts/units/items, a design engineer must ensure the following items in a system under consideration [2]:

- Availability of sufficient information in task instructions and number plate identification for allowing users to decide with confidence whether two similar parts/components are interchangeable.
- Availability of adapters to make physical interchangeability possible when total interchangeability is not practicable.
- Existence of appropriate functional interchangeability when physical interchangeability is a design characteristic.
- No change in methods of connecting and mounting when there are part or unit modifications.
- Total interchangeability of all identical parts identified as interchangeable through some appropriate identification system.
- Avoidance of differences in size, shape, mounting, etc.

Finally, it is to be noted that when functional interchangeability is not required, there is no need to have physical interchangeability.

Identification is concerned with labeling or marking of parts, controls, and test points for facilitating tasks such as repair and replacement. When parts, controls, and test points are not highlighted effectively, the performance of maintenance tasks becomes more difficult, takes longer, and increases the chances for making errors. Types of identification include equipment identification and part identification. Additional information on identification is available in Refs. [1, 2].

8.9 STANDARDIZATION AND MODULARIZATION

Standardization is the attainment of maximum practical uniformity in product design [12, 13]. More specifically, it restricts to a minimum variety of components that a product will need. There are many goals of standardization. Some of the important ones are shown in Figure 8.3 [1].

Standardization should be the main goal of design, because the use of nonstandard components may result in lower reliability and increased maintenance.

Nonetheless, past experiences of the years clearly indicate that the lack of standardization is generally due to poor communication, among design engineers, contractors, subcontractors, users, etc.

Some of the benefits of standardization are as follows [1]:

- Reduction in installation and wiring errors because of variations in characteristics of similar items.
- Reduction in the occurrence of accidents caused by incorrect or unclear procedures.
- Reduction in the danger of using the wrong parts.
- Reduction of manufacturing costs, design time, and maintenance cost and time.
- Elimination in the need for special or close-tolerance parts.
- Reduction in procurement, stocking, and training problems.
- Better reliability.

Modularization is the division of a product into functionally and physically distinct units to allow easy replacement and removal. The degree of modularization of a product is dictated by factors such as function, practicality, and cost. Some of the useful guidelines associated with designing modularized products are as follows [12, 14]:

- Aim to design modules for maximum ease of operational testing when they are removed from the actual system/equipment.
- Design control levers and linkages for allowing easy disconnection from components. This will make it easier to replace components.

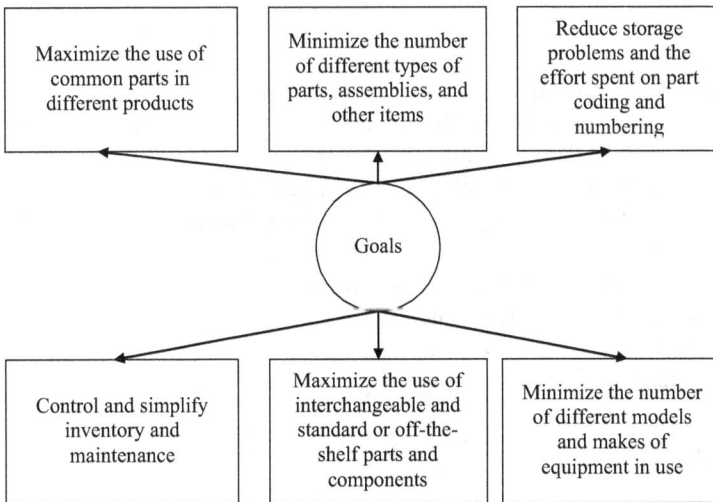

FIGURE 8.3   Some of the important standardization goals.

- Follow an integrated approach to design. More clearly, consider design, modularization, and material problems simultaneously.
- Design the system/equipment so that a single person can replace any failed part without any difficulty.
- Emphasize modularization for forward levels of maintenance as much as possible for increasing operational capability.
- Make each modular unit light and small so that a person can carry and handle it in an effective manner without any difficulty.
- Aim to make all modules and parts as uniform as possible with respect to size and shape.
- Divide the equipment/item under consideration into many modular units/parts.
- Aim to make each module capable of being inspected independently.

There are many advantages of modularization. Some of the important ones are lower levels of skill needed for replacing modular units in the field as well as the need for fewer tools; ease of dividing up maintenance responsibilities; less costly and less time consuming training of maintenance personnel; reduction in equipment downtime as recognition, isolation, and replacement of fault items become easier; relative ease of maintaining a divisible configuration; ease of modifying existing equipment with the latest functional units by simply replacing their older equivalents; simpler new equipment design; and shorter design time [12].

PROBLEMS

1. Describe the cause-and-effect diagram.
2. What are the basic steps used to perform failure modes, effects, and criticality analysis (FMECA)?
3. What is the difference between FMECA and failure mode and effects analysis (FMEA)?
4. Describe the total quality management (TQM) method.
5. What are the most frequently addressed maintainability design factors?
6. What are the factors that affect accessibility?
7. What are the basic principles of interchangeability?
8. What are the important goals of standardization and its (i.e., standardization's) benefits?
9. What are the advantages of modularization?
10. Describe the following two items:
    - Identification
    - Simplification

REFERENCES

1. Dhillon, B.S., *Engineering Maintainability*, Gulf Publishing, Houston, Texas, 1999.
2. AMCP 706-133, *Engineering Handbook: Maintainability Engineering Theory and Practice*, Department of Defense, Washington, DC, 1976.

3. Ishikawa, K., *Guide to Quality Control*, Asian Productivity Organization, Tokyo, 1976.

4. Dhillon, B.S., *Design Reliability: Fundamentals and Applications*, CRC Press, Boca Raton, Florida, 1999.

5. Bowles, J.B., Bonnell, R.D., *Failure, Mode, Effects and Criticality Analysis, in Tutorial Notes: Annual Reliability and Maintainability Symposium*, 1994, Evans Associates, Durham, NC.

6. Dhillon, B.S., *Systems Reliability, Maintainability, and Management*, Petrocelli Books, New York, 1983.

7. NUREG-0492, *Fault Tree Handbook*, U.S. Nuclear Regulatory Commission, Washington, DC, 1981.

8. Dhillon, B.S., Singh, C., *Engineering Reliability: New Techniques and Application*, John Wiley and Sons, New York, 1981.

9. Walton, M., *Deming Management at Work*, Putnam, New York, 1990.

10. Burati, J.L., Matthews, M.F., Kalidindi, S.N., Quality Management Organizations and Techniques, *Journal of Construction Engineering and Management*, Vol. 118, 1992, pp. 112–128.

11. Gevirtz, C.D., *Developing New Products with TQM*, McGraw-Hill, New York, 1994.

12. AMCP-706-134, *Engineering Design Handbook: Maintainability Guide for Design*, Department of Defense, Washington, DC, 1972.

13. Ankenbrandt, F.L., et al., *Maintainability Design*, Engineering Publishers, Elizabeth, New Jersey, 1963.

14. Rigby, L.V., et al., Guide to Integrated System Design for Maintainability, Report No. ASD-TR-61-424, U.S. Air Force Systems Command, Wright-Patterson Air Force Base, Ohio, 1961.

9 Maintainability Management and Human Factors in Maintainability

9.1 INTRODUCTION

Just like in other areas of engineering, management plays an important role in the practice of maintainability engineering. Its tasks range from simply managing maintainability personnel to effective execution of technical maintainability-related tasks. Maintainability management can be examined from different perspectives such as management of maintainability as an engineering discipline, the role maintainability plays at each phase in the life cycle of the system/product being developed, and the place of the maintainability function within the organizational structure [1].

Human factor is an important discipline of engineering and it exists because humans make errors in using and maintaining engineering systems; otherwise, it would be rather difficult to justify the discipline's existence. Although the modern history of human factors may be traced back to 1898, when Frederick W. Taylor performed various studies to determine the most suitable design of shovels, human factors have only been an important element of maintainability work since World War II [2, 3]. During this war, the performance of military equipment clearly proved that equipment is only as good as the persons operating and maintaining it. This means that humans play an important role in the overall success of a system.

This chapter presents various important aspects of maintainability management and human factors in maintainability.

9.2 MAINTAINABILITY MANAGEMENT FUNCTIONS IN THE PRODUCT LIFE CYCLE

During the product life cycle, as maintainability issues arise, various types of maintainability management-associated tasks are conducted. An effective maintainability program incorporates a dialogue between the user and manufacturer throughout the product life cycle, which can be divided into four phases as shown in Figure 9.1 [4].

Phase I is the concept development phase of the product life cycle. During this phase, the product operational-related needs are translated into a set of operational requirements and all high-risk areas are highlighted. The main maintainability management-related task during this phase is concerned with determining the product effectiveness requirements as well as determining, from the product's

DOI: 10.1201/9781003365204-9

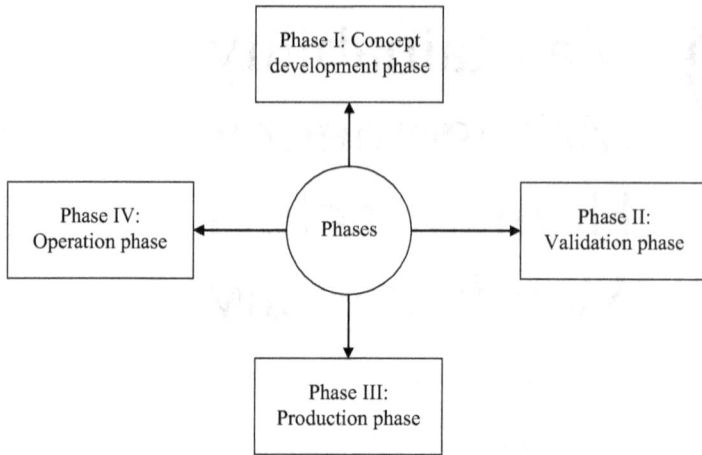

FIGURE 9.1   Product life cycle phases.

purpose and intended operation, all the required field support policies and other provisions.

Phase II is the validation phase of the product life cycle. Some of the maintainability management tasks associated with this phase are as follows [4]:

- Developing a maintainability program plan that satisfies contractual requirements.
- Developing a plan for maintainability testing and demonstration.
- Establishing maintainability incentives and penalties.
- Establishing maintainability policies and procedures for the validation phase and the subsequent full-scale engineering effort.
- Performing maintainability allocations and predictions.
- Providing assistance to maintenance engineering in areas such as conducting maintenance analysis and developing logistic policies.
- Participating in design reviews.
- Developing a planning document for data collection.
- Coordinating and monitoring maintainability efforts throughout the organization.

Phase III is the production phase of the product life cycle. Some of the maintainability management tasks associated with this phase are participating in the development of appropriate controls for errors, process variations, and other problems that may affect maintainability directly or indirectly; evaluating proposals for changes with respect to their impact on maintainability; ensuring the eradication of all shortcomings that may degrade maintainability; evaluating production test trends from the standpoint of adverse effects on maintainability requirements; and monitoring production processes [3].

Phase IV is the operation phase (i.e., the final phase) of the product life cycle. No specific maintainability management-associated tasks are involved with this phase, but the phase probably is the most significant because during this period, the product's true logistic support and cost-effectiveness are demonstrated. Thus, essential maintainability-associated data can be collected for use in future applications [3].

9.3 MAINTAINABILITY ORGANIZATION FUNCTIONS

A maintainability organization conducts a wide variety of functions that can be grouped under five distinct classifications shown in Figure 9.2 [1, 3].

The administrative classification includes functions such as developing and issuing policies and procedures for application in maintainability efforts, acting as a liaison with higher-level management and other concerned bodies, preparing budgets and schedules, taking part in program management and design reviews, monitoring the maintainability organization's output, organizing the maintainability effort, assigning maintainability-related responsibilities, and providing maintainability training [1, 5].

The design classification includes functions such as reviewing product design with respect to maintainability, providing consulting services to professionals such as design engineers, participating in the development of maintainability design criteria and guidelines, approving design drawings from the maintainability standpoint, and preparing maintainability-related design documents [1, 3].

The analysis classification includes functions such as performing analysis of maintainability data obtained from the field and other sources, participating in product engineering analysis to safeguard maintainability interests, reviewing product and system specification documents from the standpoint of maintainability requirements, conducting maintainability allocation and prediction studies, taking part in or conducting required maintenance analysis, and developing maintainability demonstration documents [1, 5].

The coordination classification includes functions such as follows [1, 3]:

- Coordinating maintainability training activities for all people involved.
- Acting as a liaison with subcontractors on maintainability-associated issues.

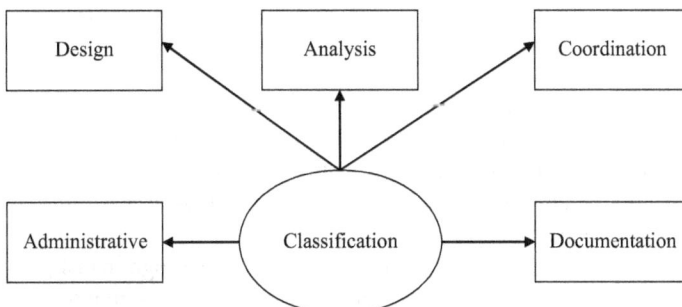

FIGURE 9.2   Maintainability organization function classifications.

- Interfacing with product engineering and other engineering disciplines.
- Coordinating with professional societies, governments, and trade associations on maintainability-associated matters.

Finally, the documentation classification includes functions such as developing and maintaining handbook data and information on maintainability-associated issues, documenting the results of maintainability analysis and trade-off studies, establishing and maintaining a library facility that contains important maintainability documents and information, documenting information concerning maintainability management, developing and maintaining a maintainability data bank, developing maintainability-related data and feedback reports, and documenting maintainability design review results [1, 3, 5].

9.4 MAINTAINABILITY PROGRAM PLAN

This is a very important document that contains maintainability-related information concerning a project under consideration. It is developed either by the system/product manufacturers or the user, depending on factors such as the philosophy of the decision-makers and the nature of the project. Eleven important elements of a maintainability program plan are as follows [3, 6]:

- **Element I: Objectives.** These are basically the descriptions of the overall requirements for the maintainability program as well as goals of the plan.
- **Element II: Policies and procedures.** The main purpose of these policies and procedures is to assure customers that the group implementing the maintainability program will conduct its assigned task in an effective manner. Under the policies and procedures, the management's overall policy directives for maintainability are also incorporated or referenced. The directives address items such as maintainability demonstration methods, methods to be employed for maintainability allocation and prediction, data collection and analysis, and participation in design reviews and evaluation.
- **Element III: Organization.** A detailed organizational breakdown of the maintainability group involved in the project is provided along with the enterprise's overall structure. In addition, information concerning the background and the maintainability group personnel's experience, a work breakdown structure, and a list of the personnel assigned to each task are provided.
- **Element IV: Maintainability design criteria.** This part of the plan discusses specific maintainability-associated design features applicable to the item under consideration. In addition, the description may relate to quantitative and qualitative factors concerning areas such as accessibility, packaging, interchangeability, or parts selection.
- **Element V: Maintainability program tasks.** Each program task, task input requirements, task schedule, expected task output results, major milestones, and projected cost are described in detail.

- **Element VI: Organizational interfaces.** This section describes the lines of communication as well as the relationships between the maintainability group and the overall organization. Some of the areas of interface are design, product engineering, reliability engineering, testing and evaluation, human factors, and logistic support, as well as customers and suppliers.
- **Element VII: Program review, evaluation, and control.** This section discusses the methods and techniques to be utilized for technical design reviews, program reviews, and feedback and control. Also, it describes a risk management plan and discusses the evaluation and incorporation of proposed changes and corrective measures to be taken in given situations.
- **Element VIII: Subcontractor and supplier activity.** This section discusses the organization's relationships with subcontractors and suppliers connected to the maintainability program. Furthermore, it outlines the procedures to be utilized for review and control within the framework of those relationships.
- **Element IX: Maintenance concept.** This section discusses basic maintenance-related requirements of the product under consideration and issues such as qualitative and quantitative objectives for maintainability and maintenance, test and support equipment criteria, organizational responsibilities, spare and repair part factors, and operational and support concepts.
- **Element X: Technical communications.** This section briefly discusses every deliverable item and their associated due dates.
- **Element XI: References.** This section lists all documents related to the maintainability requirements (e.g., plans, specifications, and applicable standards).

9.5 MAINTAINABILITY DESIGN REVIEWS

Design reviews are a critical element of modern design practices, and they are carried out during the product design phase. The primary objective of design reviews is determining the progress of the ongoing design effort as well as ensuring the application of correct design practices. The members of the design review team assess potential and existing problems in various areas concerned with the product under consideration including maintainability. Many maintainability-associated issues require careful attention during design reviews.

Some of these issues are conformance to maintainability design specifications, design constraints and specified interfaces, identified maintainability problem areas and proposed corrective actions, physical configuration and layout drawings and schematic diagrams, use of built-in monitoring and fault-isolation equipment, maintainability test data obtained from experimental models and breadboards, maintainability prediction results, maintainability trade-off study results, maintainability assessments using test data, use of on-line repair with redundancy, assessments of maintenance and supportability, maintainability demonstration test data, use of unit replacement approach, verification of maintainability design test plans, corrective actions taken and proposed, selection of parts and materials, use of automatic test equipment, and failure mode and effect analysis results [3, 7–9].

9.6   TYPICAL HUMAN BEHAVIORS

Over the years, many researchers have studied human behaviors and made conclusions about many typical behaviors. The knowledge of such behaviors can be quite useful in maintainability-related work directly or indirectly. Some of these typical behaviors are as follows [3, 10, 11]:

- People tend to use their hands for testing or examining.
- People become complacent and less careful after successfully handling hazardous objects over a lengthy period of time.
- People quite often estimate distance, speed, or clearance poorly. They generally underestimate large or horizontal distances and overestimate short distances.
- People generally conduct their assigned tasks while thinking about other things.
- People are generally reluctant to admit they do not see objects clearly, whether because of poor illumination or poor eyesight.
- People generally read labels and instructions incorrectly or overlook them altogether.
- People generally respond irrationally in emergency situations.
- People are generally too impatient to take the necessary time to observe precautions.
- After performing a procedure, people generally fail to recheck their work for errors.
- People are generally reluctant to admit mistakes.
- People get easily confused with unfamiliar things.

9.7   HUMAN BODY MEASUREMENTS AND SENSORY CAPABILITIES

In maintainability work, there is a need for an understanding of human body measurements and sensory capabilities. Both these items are discussed in the following two subsections.

9.7.1   HUMAN BODY MEASUREMENTS

This information is very important in designing for maintainability people normally operate and maintain engineering systems/products. It helps designers for ensuring that equipment and products under consideration will accommodate operating and maintenance personnel of varying weights, shapes, and sizes. In turn, these people will conduct their tasks effectively.

Normally, human body-associated requirements are outlined in the product or system design specification, particularly when the equipment is being developed for use in a military application. For example, MIL-STD-1472 [12] states, "Design shall insure operability and maintainability by at lest 90 percent of the user population" and "The design range shall include at least the 5th and 95th percentiles for design-critical body dimensions."

Furthermore, the standard states that the use of anthropometric data should take into consideration factors such as the position of the body during task performance, the nature and frequency of tasks to be conducted, the increments in the design-critical dimensions imposed by protective garments, the difficulties associated with intended tasks, the need to compensate for obstacles, the mobility and flexibility requirements of the task, and so on.

Some body-associated dimensions of the U.S. adult population (18–79 years) are presented in Table 9.1 [11–13].

Some of the useful pointer for engineering designers concerning the application of body strength and force are as follows [14]:

- The maximum handgrip strength of a 25-year-old male is around 125 pounds.
- With the use of the whole arm and shoulder, the maximum exertable force is increased.
- The degree of force that can be exerted is determined by factors such as body position, body parts involved, the object involved, and direction of force applied.
- The maximum push force for side-to-side motion is around 90 pounds.
- A person's arm strength reaches its peak around age 25 years.
- Pull force is greater from a sitting than from a standing position.

9.7.2 Human Sensory Capabilities

In maintainability work, there is a definite need for a good understanding of human sensory capabilities, as they apply to areas such as noise, parts identification, and color coding. The five major senses possessed by humans are hearing, sight, touch, smell, and taste. Humans can also sense items such temperature, vibration, pressure, acceleration (shock), and linear motion.

TABLE 9.1
Some Body-Associated Dimensions of the U.S. Adult Population (18–79 Years)

No.	Description	5th Percentile (in inches)		95th Percentile (in inches)	
		Male	Female	Male	Female
1	Weight	126 (lb)	104 (lb)	217 (lb)	199 (lb)
2	Standing height	63.6	59	72.8	67.1
3	Seated width	12.2	12.3	15.9	17.1
4	Seated eye height	28.4	27.4	33.5	31
5	Sitting height	33.2	30.9	38	35.7

The first three of the major senses are described in the following three subsections and the information on other sensors is available in Refs. [1, 3, 7].

9.7.2.1 Hearing

This sensor is an important factor in maintainability work, as excessive noise may lead to problems such as reduction in the workers' efficiency, loss in hearing if exposed for long periods, adverse effects on tasks, and need for intense concentration or a high degree of muscular coordination. In order to lower the effects of noise, some useful guidelines related to maintainability are as follows [7]:

- Prevent unprotected repair personnel from entering areas with sound levels greater than 150 dB.
- Incorporate into the equipment appropriate acoustical design and mufflers and other sound-proofing devices in areas where maintenance tasks must be conducted in the presence of extreme noise.
- Protect maintenance personnel by issuing protective devices where noise reduction is not possible.
- Keep noise levels below 85 dB in areas where the presence of maintenance personnel is necessary.

9.7.2.2 Sight

This is another sensor that plays an important role in maintainability-related work. Sight is stimulated by electromagnetic radiation of certain wavelengths, often referred to as the visible segment of the electromagnetic spectrum. In daylight, the human eye is quite sensitive to greenish-yellow light, and it sees differently different angles.

Some of the important factors concerning color in regard to the human eye are as follows:

- The color reversal phenomenon may take place when one is staring, for example, at a green or red light and then glances away. In such situations, the signal to the brain may reverse the color.
- In poorly lit areas or at night, it may be impossible to determine at a distance, the color of a small point source of light (e.g., a small warning light). In fact, the light colors will appear to be white.
- Generally, the eye can perceive all colors when looking straight ahead. However, with an increase in viewing angle, color perception decreases significantly.

Some useful guidelines for designers and others are choose colors in such a way that color-weak individuals do not get confused, use red filters with a wavelength greater than 6500 A, and avoid placing too much reliance on color when critical tasks are to be conducted by fatigued persons [15].

9.7.2.3 Touch

This complements human ability for interpreting visual and auditory stimuli. In maintainability work, the touch sensor may be used for relieving eyes and ears part of the load. For example, its application could be the recognition of control knob shapes with or without using other sensors.

The touch sensor's use in technical work is not new; it has been used for many centuries by craft workers for detecting surface roughness and irregularities. Furthermore, as per Ref. [16], the detection accuracy of surface-related irregularities dramatically improves when the worker moves an intermediate piece of paper or thin cloth over the object surface rather than simply using his/her bare fingers.

9.8 AUDITORY AND VISUAL WARNINGS IN MAINTENANCE-RELATED WORK

In maintenance-related work, various auditory and visual warning devices are used for the safety of maintenance personnel. A clear understanding of such devices is very important. Some examples of warning devices used in maintenance-related work are bells, sirens, and buzzers.

In maintainability design, with respect to the use of auditory warning devices, attention should be given to factors such as follows [3]:

- Easy detectability.
- Suitability for getting the attention of repair personnel.
- Noncontinuous and high-pitched tones above 2000 Hz.
- Use of warbling or undulating tones and sound at least 20 dB above threshold level.
- Distinctiveness.
- No requirement for interpretation when maintenance personnel are conducting repetitive tasks.

Additional design recommendations for auditory warning devices to address corresponding conditions (in parentheses) are use low frequencies (sound is expected to pass through partitions and bend around obstacles), use manual shut-off mechanism (warning signal must be acknowledged), select a frequency that makes the signal audible through other noise (presence of background noise), use high intensities and avoid high frequencies (repair personnel are conducting their tasks far from the signal source), and modulate the signal to generate intermittent beeps (signal must command maintenance person's attention).

Some of the conditions for using auditory presentation are that the maintenance person is moving around continuously, the message is simple, the message requires immediate action, the maintenance person is overburdened with visual stimuli, the message receiving location is too brightly lit, and the message is short. Similarly, some of the conditions for using visual presentation are that the maintenance person is overburdened with auditory stimuli, the maintenance person's job allows him/her to remain in one place, the message receiving location is too noisy, the message is complex, the message does not require immediate action, and the message is long.

Three situations that require the simultaneous use of both auditory and visual signals are shown in Figure 9.3 [7].

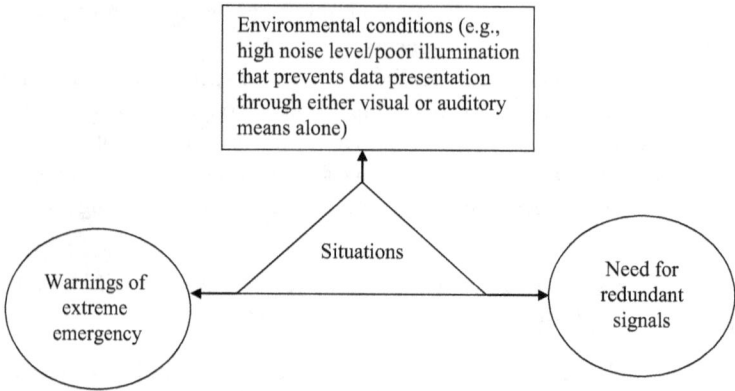

FIGURE 9.3   Situations that require the simultaneous use of both auditory and visual signals.

9.9   HUMAN FACTORS-RELATED FORMULAS

Over the years, human factors researchers have developed many mathematical formulas for estimating human factors-related information. This section presents some of these formulas considered quite useful for maintainability work.

9.9.1   GLARE CONSTANT ESTIMATION

For various maintainability-associated tasks, glare can be a serious problem. The glare constant's value can be calculated by using the following equation [17]:

$$\theta_g = \frac{(LS)^{1.6}(SA)^{0.8}}{(AVGD)^2(GBL)} \tag{9.1}$$

where

θ_g is the glare constant value.
LS is the source luminance.
SA is the solid angle subtended at the eye by the source.
$AVGD$ is the angle between the viewing direction and the glare source direction.
GBL is the general background luminance.

9.9.2   THE DECIBEL

The sound intensity level is measured in terms of decibels, the basic unit named after Alexander Graham Bell (1847–1922), the inventor of telephone. The sound-pressure level, in decibels, is expressed by [18, 19]:

$$SPL = 10\log_{10}\left(\frac{P^2}{P_s^2}\right) \tag{9.2}$$

where

SPL is the sound-pressure level expressed in decibels.
P^2 is the sound pressure, squared, of the sound to be measured.
P_s^2 is the standard reference sound pressure, representing zero decibels, squared.
P_s is the faintest 1000 Hz tone that an average young person can hear (i.e., under normal circumstances).

9.9.3 LIFTING LOAD ESTIMATION

This formula is concerned with estimating the maximum lifting load for an individual. This information could be quite useful in regard to structuring various maintenance-related tasks. The maximum lifting load is defined by [20],

$$LL_m = c\left(IBMS\right) \qquad (9.3)$$

where

LL_m is the maximum lifting load for a person.
c is a constant whose values are 1.1 and 0.95 for males and females, respectively.
$IBMS$ is the isometric back muscle strength of the person.

9.9.4 CHARACTER HEIGHT ESTIMATION FORMULAS

This section presents two such formulas.

9.9.4.1 Formula I

This formula was developed in 1959 by Peters and Adams for determining character height by taking into consideration factors such as viewing distance, illumination, viewing conditions, and importance of reading accuracy [21]. Thus, the character height is expressed by

$$C_h = \lambda V_d + \theta_{im} + CF_{vci} \qquad (9.4)$$

where

C_h is the character height in inches.
λ is a constant whose specified value is 0.0022.
V_d is the viewing distance expressed in inches.
θ_{im} is the correction factor associated with importance. Its stated value for important items such emergency labels is 0.075 and for other items is $\theta_{im} = 0$.
CF_{vci} is the correction factor for viewing conditions and illuminations. Its recommended values for various corresponding viewing conditions and illuminations (in parentheses) are 0.26 (unfavorable reading conditions, below 1

foot-candle), 0.16 (favorable reading conditions, below 1 foot-candle), 0.16 (unfavorable reading conditions, above 1 foot-candle), and 0.06 (favorable reading conditions, above 1 foot-candle).

Example 9.1

Assume that the viewing distance of an instrument panel is estimated to be approximately 45 inches. Calculate the height of the characters that should be used on the panel for $CF_{vci} = 0.26$ and $\theta_{im} = 0.075$. By substituting the above given values into Equation (9.4), we obtain

$$C_h = (0.0022)(45) + 0.075 + 0.26$$
$$= 0.434$$

Thus, the height of the characters should be 0.434 inches.

9.9.4.2 Formula II

Normally, for a comfortable arm reach for conducting control and adjustment-oriented tasks, the instrument panels are installed at a viewing distance of 28 inches. Thus, marking, letter, and number sizes are based on this viewing distance. However, sometimes, the need may arise to vary this distance; under such conditions, the following equation can be used for estimating the required character height [19, 22]:

$$CH = VD_s H_s / 28 \qquad\qquad (9.5)$$

where

CH is the character height estimate at the specified viewing distance (VD_s) expressed in inches.

H_s is the standard or recommended character height at a viewing distance of 28 inches.

Example 9.2

Assume that a meter has to be read at a distance of 60 inches. The recommended numeral height at a viewing distance of 28 inches at low luminance is 0.31 inches. Calculate the numeral height for the viewing distance of 60 inches.

By substituting the given data values into Equation (9.5), we obtain

$$CH = (60)(0.31)/28$$
$$= 0.66 \text{ Inches}$$

Thus, the estimate for the numeral height for the stated viewing distance is 0.66 inches.

PROBLEMS

1. Discuss maintainability management functions during the following two phases of the product life cycle:
   - Concept development
   - Validation
2. Discuss maintainability organization functions.
3. What are the important elements of a maintainability program plan? List at least 11 of them.
4. List at least 13 maintainability-related issues that require careful attention during product design reviews.
5. List at least ten typical human behaviors.
6. List at least six useful pointers for engineering designers concerning the application of body force and strength.
7. What are the human sensory capabilities? Discuss at least two such capabilities in detail.
8. What are the factors to which attention should be given in maintainability design with respect to the use of auditory warning devices?
9. Write down the formula to estimate the maximum lifting load for a person.
10. Write down the formula to estimate the glare constant value.

REFERENCES

1. AMCP-706-133, *Engineering Design Handbook: Maintainability Engineering Theory and Practice*, Department of Defense, Washington, DC, 1976.
2. Chapanis, A., *Man-Machine Engineering*, Wadsworth, Belmont, California, 1965.
3. Dhillon, B.S., *Engineering Maintainability*, Gulf Publishing, Houston, Texas, 1999.
4. Dhillon, B.S., Reiche, H., *Reliability and Maintainability Management*, Van Nostrand Reinhold, New York, 1985.
5. MIL-STD-470, *Maintainability Program Requirements for Systems and Equipment*, Department of Defense, Washington, D.C., 1966.
6. Blanchard, B.S., Verma, D., Peterson, E.L., *Maintainability: A Key to Effective Serviceability and Maintenance Management*, John Wiley and Sons, New York, 1995.
7. AMCP-706-134, *Engineering Design Handbook: Maintainability Guide for Design*, Department of Defense, Washington, DC, 1972.
8. Patton, J.D., *Maintainability and Maintenance Management*, Instrument Society of America, Research Triangle Park, NC, 1980.
9. Pecht, M., Ed., *Product Reliability, Maintainability, and Supportability Handbook*, CRC Press, Boca Raton, Florida, 1995.
10. Nertney, R.J., Bullock, M.G., Human Factors in Design, Report No. ERDA-76-45-2, The Energy Research and Development Administration, U.S. Department of Energy, Washington, DC, 1976.
11. Woodson, W.E., *Human Factors Design Handbook*, McGraw-Hill, New York, 1981.
12. MIL-STD-1472, *Human Engineering Design for Military Systems, Equipment, and Facilities*, Department of Defense, Washington, DC, 1972.
13. Dhillon, B.S., *Advanced Design Concepts for Engineers*, Technomic, Lancaster, PA, 1998.
14. Henney, K., Ed., *Reliability Factors for Ground Electronic Equipment*, The Rome Air Development Center, Griffis Air Force Vase, Rome, NY, 1955.

15. Woodson, W., *Human Engineering Suggestions for Designers of Electronic Equipment, in NEL Reliability Design Handbook, U.S.* Naval Electronics Laboratory, San Diego, CA, 1955, 12.1–12.5.

16. Lederman, S., Heightening Tactile Impressions of Surface Texture, in *Touch*, Gordon, G., Ed., Pergamon Press, Elmsford, New York, 1978, pp. 20–32.

17. Oborne, D.J., *Ergonomics at Work*, John Wiley and Sons, New York, 1982.

18. Adams, J.A., *Human Factors Engineering*, MacMillan, New York, 1989.

19. McCormick, E.J., Sanders, M.S., *Human Factors in Engineering and Design*, McGraw Hill, New York, 1982.

20. Poulsen, E., Jorgensen, C., Back Muscle Strength, Lifting and Stooped Working Postures, *Applied Ergonomics*, Vol. 2, 1971, pp. 133–137.

21. Peters, G.A., Adams, B.B., Three Criteria for Readable Panel Markings, *Product Engineering*, Vol. 30, No. 2, 1959, pp. 55–57.

22. Dale Huchingson, R., *New Horizons for Human Factors in Design*, McGraw Hill, New York, 1981.

10 Maintainability Testing and Demonstration

10.1 INTRODUCTION

The primary function of maintainability testing and demonstration is verifying the maintainability features that been designed and built into a product/system [1, 2]. Testing and demonstration also provide the customer with confidence; prior to making production commitments, the equipment design under consideration meets the maintainability-related requirements. Prior to the testing and demonstration phase, the maintainability program's tasks have been basically analytical. They have provided a certain degree of assurance through steps such as conducting allocations and predictions, developing design criteria, and participating in design reviews that both the qualitative and quantitative maintainability-related requirements would be satisfied [3]. However, the main drawback of these evaluations is that they do not reflect practical experience with the actual hardware.

Thus, it is very important to add realistic evaluations to analytical evaluations by carrying out real maintainability tests and demonstrations with the equipment in its operational environment. This chapter presents various important aspects of maintainability testing and demonstration.

10.2 PLANNING AND CONTROL REQUIREMENTS FOR MAINTAINABILITY TESTING AND DEMONSTRATION

In order to gain maximum benefits from maintainability tests and demonstrations requires careful planning and control. Thus, the requirements for planning and control are divided into six categories as shown in Figure 10.1 [3]. Each of these categories are described in the following six subsections:

10.2.1 CATEGORY I: DEVELOPING A DEMONSTRATION PLAN

A good demonstration plan should cover areas such as test planning, administration, and control; test conditions; and test documentation, analysis, and reporting. The plan should also conform to the specifics described in MIL-STD-471 [4].

During initial contractor/manufacturer participation in a program such as the validation phase, the first step is to conceive, propose, and negotiate the demonstration test planning subject. As the program progresses, the mutually agreed-upon test plans are updated in regard to schedules, personnel selection, demonstration model designation, and identification of logistic support resource-related requirements. It is to be noted that the important factor in accomplishing the demonstration on schedule and

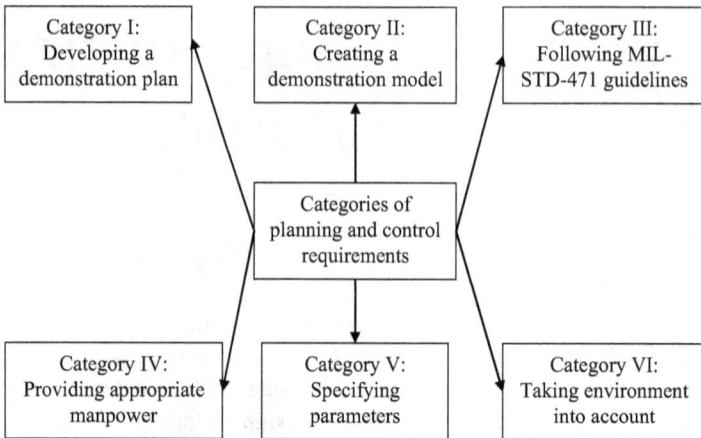

FIGURE 10.1 Categories of planning and control requirements for maintainability testing and demonstration.

within budget is administration and control of the demonstration. Some of the elements of administration and control are as follows:

- Method of organization.
- A team approach if desired.
- Test monitoring.
- Organizational interfaces.
- Assignment of responsibilities.
- Test event scheduling.
- Cost control.
- Test data collection, reporting, and analysis.

The type and complexity of the equipment under consideration plays a very important role in shaping the requirement for test documentation. Documentation requirements generally include items such as failure reports, task selection work sheets, frequency and distribution work sheets, demonstration work sheets, demonstration task data sheets, interim demonstration reports, demonstration analysis work sheets, and final reports.

10.2.2 Category II: Creating a Demonstration Model

The ideal hardware configuration for the formal test and demonstration, which will be the basis for the decision to accept or reject, would be a configuration identical to that of the final system/product. However, in reality, this may not be always possible. The demonstration can use instead a prototype model that incorporates improvements in design, costs, and scheduling made for rectifying visible shortcomings and safety hazards. The main drawback of using prototype models is that they largely consist of handcrafted parts that nearly fit together versus the mass-produced parts that will be used in the end product/system. Difficulties, with tolerances and other quality control problems with the use of mass-produced parts, will therefore not show up during the testing and demonstration process.

A mock-up model submitted by the manufacturer/contractor can also effectively demonstrate maintainability-related features. Such models serve the following two basic functions:

- **Function I:** Providing a basic mechanism for demonstrating the product's proposed quantitative parameters and qualitative design-related features for maintainability.
- **Function II:** Providing a designer's tool for visibility, experimentation, packaging limitation, and planning, prior to release of the final design or drawing.

10.2.3   CATEGORY III: FOLLOWING MIL-STD-471 GUIDELINES

This 1966 U.S. Department of Defense document entitled "Maintainability Program Requirements" lays out guidelines that manufacturers should carefully consider in the planning and control of maintainability demonstrations [4]. The topics covered in the document include test conditions, selecting a test method, establishing test teams, and suggested test support materials; the pre-demonstration, formal demonstration, phases; administration, control, reporting, evaluation, and analysis procedures for the demonstration; selection, performance, and sampling of corrective and preventive maintenance tasks; and data collection. Finally, it is to be noted that the MIL-STD-471 guidelines are considered essential for an effective maintainability demonstration.

10.2.4   CATEGORY IV: PROVIDING APPROPRIATE MANPOWER

The people who conduct maintainability demonstrations will be essential to their success, and it is very important that they clearly possess backgrounds and skill levels similar to those of the product's final user, maintenance, and operating personnel. One way to do this is to have such personnel from the client organization conduct the test. Finally, it is to be noted that Ref. [3] provides useful guidelines for selecting demonstrators.

10.2.5   CATEGORY V: SPECIFYING PARAMETERS

The primary purpose of a formal maintainability demonstration process is verifying compliance with specified parameters. The specifications for demonstration parameters should be stated in quantitative terms. Some examples of measurable time parameters are mean preventive maintenance time, mean corrective maintenance time, and mean time to repair (MTTR).

10.2.6 CATEGORY VI: TAKING ENVIRONMENT INTO ACCOUNT

Past experiences, over the years, have clearly shown that equipment downtime may vary significantly between laboratory-controlled conditions and actual operational conditions. Environment is therefore a very important factor in testing, and those responsible must carefully consider factors such as test facilities, limitation simulations, and support resource needs.

10.3   MAINTAINABILITY TEST APPROACHES

Not all maintainability-related tests are formal accept/reject demonstration tests. In fact, there are many points in the product life cycle as well as in related maintainability program tasks that require test data, both before and after the formal decision to accept or reject. Test data may be necessary for administrative and logistic control for updating corrective actions or modifications, for making decisions about maintainability design requirements, or for evaluating life cycle maintenance support. The maintainability test approaches that can provide this data fall into the following six categories [2, 3]:

- **Category I: Functional tests.** These tests closely simulate normal operating conditions for establishing the product's state of readiness to conduct its proposed mission effectively. The functional tests can proceed on a system-wide level or focus on components such as replaceable subassemblies. These tests are conducted and required at each point of evaluation for the product.
- **Category II: Marginal tests.** The purpose of applying these tests is to isolate potential problems through the simulation of abnormal operating conditions. Such tests supply unrelated stimuli to the system/equipment under conditions such as extreme heat, vibrations, and lowered power supply voltages. Marginal testing provides greatest value as part of fault prediction, where it highlights various incipient failures resulting from abnormal environments and operating conditions.
- **Category III: Open-loop tests.** These tests represent a refinement of dynamic and static tests. They do not provide intelligence feedback to the item being tested, that is, the stimulus is not adjusted. Open-loop tests generally provide better maintenance-related information than closed-loop tests because they make a direct observation of the system transfer function without the modifying influence of feedback. This method is also cheaper and simpler than closed-loop testing. It is probably the most appropriate type of testing for maintenance purposes, because it illuminates the possibility of test instability.
- **Category IV: Dynamic tests.** These tests simulate typical operation or uses of equipment/system so that every item involved can be checked. The tests involve a continuous input signal and analysis of the corresponding output signals for determining whether system needs are fully satisfied. Dynamic tests also provide additional information on matters such as frequency responses, phase characteristics, and integration rates.

- **Category V: Static tests.** These tests are easy to carry out and provide information on the transient behavior of the item being tested. A series of intermittent, sequences input signals feed into the item and, to measure its operation, the test monitors output response signals. Static tests usually establish a confidence factor, but their use does not go beyond that.
- **Category VI: Closed-loop tests.** These tests generate information for use in evaluating design effectiveness, performance, tolerance adequacy, and other important issues. In these tests, the stimulus is adjusted on a continuous basis according to performance of the system/equipment under the test conditions. Closed-loop tests are extremely useful in situations when a high degree of accuracy is necessary and when test points radiate performance-degrading noise levels. However, closed-loop circuits/paths in equipment/system design are quite difficult to maintain and failures in the loop are difficult to diagnose.

10.4   TESTING METHODS

Maintainability demonstration determines whether a manufacturer or a development program has effectively satisfied qualitative and quantitative maintainability-related requirements. A successful maintainability demonstration depends on factors such as quality of written maintenance manuals, quality of training or repair technicians, and the quality of product design for testability. A maintainability test will not necessarily show that maintainability requirements have been met, but it does focus the manufacturer's attention on the need for meeting those requirements.

MIL-STD-471 [4] is a document widely used in carrying out maintainability demonstrations and presents many test methods. It describes policies and procedures for carrying out maintainability demonstrations at specified points during the product development life cycle. As a single maintainability parameter can seldom address all desirable maintainability characteristics, the 11 different test methods presented in the following subsections are considered quite useful for addressing many diverse maintainability parameters [2, 4, 5]. It is to be noted that ref. [4] gives the statistical aspect of the various methods presented below.

10.4.1   THE MEAN METHOD

This method is useful when the need or requirement is stated in terms of a mean value and there is a corresponding design goal value. The test plan contains two categories, Test Plan X and Test Plan Y. Test Plan X assumes lognormal distribution to determine the sample size. It also assumes that a lognormal distribution can satisfactorily represent the maintenance times, and that the variance of the logarithms of the maintenance times is already known. Test Plan Y makes no such assumptions. On the other hand, both are fixed sample tests, and a minimum sample size of 20, which makes use of the Central Limit Theorem and the asymptotic normality of the sample mean in their development.

10.4.2 THE CRITICAL MAINTENANCE TIME OR MAN-HOURS METHOD

This method is applicable when the need is stated in the following two terms:

- Required critical maintenance time or critical man-hours.
- A corresponding design goal value.

This test is distribution-free and can be used for establishing a critical upper limit on the time or man-hours required to conduct certain maintenance tasks. The following two factors are associated with this method:

- There is no need for assuming the distribution of maintenance time or man-hours.
- Both the null and alternate hypotheses refer to a fixed time and the percentile varies.

10.4.3 THE MEAN MAINTENANCE TIME AND MAXIMUM MAINTENANCE TIME METHOD

This method demonstrates maintainability indexes such as follows:

- Mean corrective maintenance time
- Mean preventive maintenance time
- Mean maintenance time, including preventive and corrective maintenance actions

For demonstrating mean corrective maintenance time, this method's procedures are based on the Central Limit Theorem. Information on the variance of maintenance times is not required. This allows the method to be used with any underlying probability distribution, provided the sample size is at least 30. The maximum maintenance time demonstration/procedure is valid for cases with lognormal underlying probability distribution of corrective maintenance task times.

10.4.4 THE CRITICAL PERCENTILE METHOD

Maintainability demonstrations employ this method when the requirement is defined in terms of the following two items:

- A required critical percentile
- A corresponding design goal value

It is to be noted that if the critical percentile is fixed at 50%, then this test method is called the test of a median. The basis for the decision criteria is the asymptotic normality of the maximum likelihood estimate of the percentile value. The method is based on the following two assumptions:

- A lognormal distribution satisfactorily describes the distribution of maintenance times.
- The variance of the logarithms of the maintenance times is already known.

10.4.5 THE MAN-HOUR RATE METHOD

This method demonstrates man-hour rates, specifically man-hours per flight hour. It uses the total accumulative flight hours and the determination, during Phase II test operation, of the total accumulative chargeable maintenance man-hours. In using this test method, the following two actions are required:

- Develop appropriate ratios of equipment operating time to flight time.
- Ensure that the predicted man-hour rate pertains to flight time rather than the equipment operating time.

10.4.6 THE PERCENTILES AND MAINTENANCE METHOD

This method uses a test of proportion for demonstrating fulfillment of maximum corrective maintenance task time at the 95th percentile, maximum preventive maintenance task time at any percentile, median corrective maintenance task time, and median preventive maintenance task time, when corrective and preventive maintenance repair time probability distributions are unknown. The following two factors are associated with the method:

- **Factor I:** A minimum sample size of 50 tasks is needed.
- **Factor II:** The plan holds the confidence level at 75% or 90%.

10.4.7 THE MAN-HOUR RATE (USING SIMULATED FAULTS) METHOD

This method demonstrates man-hour rates or man-hours per operating hour and is based on the following three items:

- Total accumulative simulated demonstration operating hours.
- Total accumulative chargeable maintenance man-hours.
- Predicted equipment failure rate.

10.4.8 THE CHARGEABLE MAINTENANCE DOWNTIME PER FLIGHT METHOD

This method, used in testing aircraft, makes use of the Central Limit Theorem. The chargeable downtime per flight is the allowable time, expressed in hours, for conducting maintenance assuming that there is a specific availability and operation readiness requirement for the aircraft.

10.4.9 THE PREVENTIVE MAINTENANCE TIMES METHOD

This method is quite useful when the stated index involves mean preventive task time and/or maximum preventive maintenance task time at any percentile, and when all possible preventive maintenance tasks need to be accomplished. The test requires no allowance for assumed probability distribution.

10.4.10 THE MEDIAN EQUIPMENT REPAIR TIME METHOD

Maintainability demonstrations employ this method when the need or the requirement is expressed in terms of an equipment repair time median. The method is based on lognormally distributed corrective maintenance task times and a sample size of 20.

10.4.11 THE COMBINED MEAN/PERCENTILE REQUIREMENT METHOD

This method is quite useful when the specification is expressed as a dual requirement for the mean and for either the 90th or 95th percentile of maintenance times, when maintenance time is lognormally distributed.

10.5 PREPARING FOR MAINTAINABILITY DEMONSTRATIONS AND EVALUATING THE RESULTS

There are many steps associated with the preparation for performing maintainability demonstrations and evaluating their results. These steps are as follows [6]:

- **Step 1: Choose the method or methods to be used.** This step is the choice of specific methods outlines in MIL-STD-471 A [4]. Selection depends on factors such as product characteristics and the parameters to be demonstrated.
- **Step 2: Develop accept/reject criteria and retest procedures in case the "accept" criteria is not met.** This step establishes accept/reject criteria and retest procedures in the event that the "accept" criteria is not met.
- **Step 3: Develop a maintainability demonstration plan and test procedures in detail.** This step is to develop in detail a maintainability demonstration plan and test procedures. The plan addresses issues such as manpower requirements, training needs, facility needs, and documentation and equipment required.
- **Step 4: Choose the maintenance task/population that is to be used for selecting the maintainability test sample.** This step deals with selecting maintenance task population out of which the maintainability test sample will be taken.
- **Step 5: Perform pretest preparation.** This step is pretest preparation. This includes preparing the facilities needed for the test and assembling the test hardware, test support equipment, documentation, and other requirements.
- **Step 6: Perform the maintainability demonstration test or tests.** This step is the performance of the maintainability test or tests.

- **Step 7: Carry out appropriate post-test tasks.** This step involves post-test tasks, such as restoring test hardware to its original form, verifying the test hardware's acceptability for use on production items, and returning test equipment and associated facilities to pretest form.
- **Step 8: Perform analysis of the test data.** This step is the analysis of test data, which includes determining whether acceptance criteria were met and analyzing the maintenance weaknesses and strengths of the product.
- **Step 9: Recommend corrective measures and retest if necessary.** In this step, corrective measures are recommended as appropriate.
- **Step 10: Prepare appropriate demonstration test documentation.** This step deals with preparing the documentation related to the demonstration test.

To avoid pitfalls in maintainability testing, the following actions are considered necessary [7]:

- Tailor MIL-STD-471 [4] for the program and product under consideration rather than relying on it entirely.
- Perform a new and different trial for every trial that highlights a deficiency.
- Conduct some "dry run" testing, if feasible.
- Limit the allowable trial repetitions as a requirement for cancelling the test, progressing into an "evaluate and fix" phase, and then repeating the test with newly stated faults.
- Improve the technical manual verification and validation process prior to the maintainability demonstration test.
- Clearly define, correct, and verify all discovered deficiencies and the related needs for corrective action.

10.6 CHECKLISTS FOR MAINTAINABILITY DEMONSTRATION PLANS, PROCEDURES, AND REPORTS

Checklists play a very important role in maintainability demonstration. The checklist for maintainability demonstration plans and procedures should cover the following items [8]:

- **Purpose and Scope.** This is a statement of general test objectives and a general description of the test to be conducted.
- **Test facilities.** Under this item is information such as a description of the test item's configuration, a general description of test facility, identification of the test locating, test area security measures, and test safety features.
- **Test requirements.** They include items such as the method of generating a candidate fault list, the method of selecting and applying faults from the candidate list, the levels of maintenance to be demonstrated, a list and schedule of test reports to be issued, and support material requirements.
- **Test participation.** Decisions to be made concerning test participation are the test team members, their assignments, and test decision-making authority.

- **Test monitoring.** This is the method of monitoring and recording test results.
- **Test schedule.** It should include three items: the start date, the finish date, and the test program review schedule.
- **Test conditions.** Two components of these conditions are the modes of equipment operation during testing, and a description of the environmental conditions under which the test will be conducted.
- **Test ground rules.** Under this item should be a list of items to which the rules apply. These items include maintenance inspection, instrumentation failures, maintenance due to secondary failures, maintenance time limits, and technical manual usage and adequacy.
- **Testability demonstration considerations.** Among the components of the testability considerations are the built-in test requirements to be demonstrated, the method of selecting and simulating candidate faults, the repair levels for which requirements will be demonstrated, and acceptable levels of ambiguity at each repair level.
- **Reference documents.** The checklist should also detail all applicable reference documents.

Finally, it is to be noted that the maintainability demonstration reports checklist include items related to test results, such as maintenance tasks planned, maintenance tasks selected, the selection method, measured repair times, data analysis calculation, qualifications of the personnel carrying out tasks, application of the accept/reject criteria, the documentation used during maintenance, and a discussion of deficiencies highlighted during testing [8].

PROBLEMS

1. List the categories of planning and control for maintainability testing and demonstration. Describe at least two of these categories.
2. Describe the following two maintainability test approaches:
   - Dynamic tests
   - Static tests
3. Describe the following two test methods presented in MIL-STD-471:
   - The mean method
   - The critical maintenance time or man-hours method
4. List at least ten test methods presented in MIL-STD-471.
5. Describe the steps associated with the preparation for performing maintainability demonstrations and evaluating their results.
6. Discuss the actions to avoid pitfalls in maintainability testing.
7. List the items the checklist for maintainability demonstration plans and procedures should cover.
8. Describe the following two maintainability test approaches:
   - Open-loop tests
   - Closed-loop tests

9. Describe the following test method presented in MIL-STD-471:
   • The mean maintenance time and maximum maintenance time method.
10. Compare the test methods: preventive maintenance times method with the median equipment repair time method presented in MIL-STD-471.

REFERENCES

1. MIL-STD-470, *Maintainability Program Requirements*, Department of Defense, Washington, D.C., 1966.
2. Dhillon, B.S., *Engineering Maintainability*, Gulf Publishing Company, Houston, Texas, 1999.
3. AMCP-706-133, *Maintainability Engineering Theory and Practice*, Department of Defense, Washington, D.C., 1976.
4. MIL-STD-471, *Maintainability Verification/Demonstration/Evaluation*, Department of Defense, Washington, D.C., 1966, (Revision "A", 1973).
5. PRIM-1, *A Primer for DOD Reliability, Maintainability, and Safety Standards*, Reliability Analysis Center, Rome Air Development Center, Griffiss Air Force Base, New York, 1988.
6. Grant Ireson, W., Coombs, C.F., Moss, R.Y., *Handbook of Reliability Engineering and Management*, McGraw-Hill Inc., New York, 1996.
7. Bentz, R.W., Pittfalls to Avoid in Maintainability Testing, *Proceedings of the Annual Reliability and Maintainability Symposium*, 1982, pp. 278–282.
8. RADC Reliability Engineer's Toolkit, *Systems Reliability and Engineering Division*, Rome Air Development Center, Griffiss Air Force Base, New York, 1988.

11 Introduction to Engineering Maintenance

11.1 NEED FOR MAINTENANCE

Each year, a vast sum of money is spent on engineering equipment maintenance worldwide, and today's maintenance-related practices are market driven, in particular for the manufacturing and process industry, service suppliers, etc. [1]. Because of this, there is a definite need for effective asset management and maintenance-related practices that can, directly or indirectly, positively influence success factors such as safety, price, quality, reliable delivery, speed of innovation, and profitability.

In the future, engineering equipment will be even more complex and computerized. It means that further computerization of equipment will increase the importance of software maintenance quite significantly, approaching, if not equaling hardware maintenance. In addition, factors such as increased computerization and complexity will result in greater emphasis on maintenance-related activities in regard to areas such as safety, human factors, cost-effectiveness, and quality [2]. In the future, creative thinking and new strategies will definitely be needed to realize all potential benefits and turn them into profitability.

11.2 ENGINEERING MAINTENANCE FACTS AND FIGURES

Some of the important facts and figures directly or indirectly concerning engineering maintenance are as follows:

- Each year, over $300 billion are spent on plant maintenance and operations by the U.S. industry [3].
- For fiscal year 1997, the request of the U.S. Department of Defense for their operation and maintenance budget was $79 billion [4].
- Each year, the U.S. Department of Defense spends around $12 billion on depot maintenance of weapon systems and equipment [5].
- Over the years, the size of a plant maintenance group in a manufacturing company has varied from 5% to 10% of the total operating force (1 to 17 persons in 1969 and 1 to 12 persons in 1981) [6].
- In 1970, a British Ministry of Technology Working Party document reported that the UK annual maintenance cost was approximately 3000 million pounds [7, 8].
- It is estimated that the cost of maintaining a military jet aircraft is around $1.6 million per year, and approximately 11% of the operating cost for an aircraft accounts for maintenance-related activities [9].

11.3   MAINTENANCE ENGINEERING OBJECTIVES

There are many objectives of maintenance engineering. Eight of the important ones are as follows [10]:

- **Objective I:** Improve and ensure maximum utilization of all maintenance facilities.
- **Objective II:** Reduce the amount and frequency of maintenance.
- **Objective III:** Establish optimum frequency and extent of preventive maintenance to be conducted.
- **Objective IV:** Improve maintenance operations.
- **Objective V:** Reduce the amount of supply support needed.
- **Objective VI:** Improve the maintenance organization.
- **Objective VII:** Lower the maintenance skills required.
- **Objective VIII:** Reduce the effect of complexity.

11.4   MAINTENANCE-RELATED MEASURES

Over the years, many indexes for measuring maintenance activity performance have been developed. Generally, the values of these indexes are calculated periodically for monitoring their trends or comparing them with established standard values. This section presents nine of these indexes [6, 11–13].

11.4.1   INDEX I

This index is useful for measuring inspection effectiveness and is defined by

$$\theta_{jc} = \frac{NJRI_t}{NIC_t} \tag{11.1}$$

where

θ_{jc} is the index parameter.
$NJRI_t$ is the total number of jobs resulting from inspections.
NIC_t is the total number of inspections completed.

11.4.2   INDEX II

This index relates the maintenance cost to the total sales revenue and is expressed as

$$\theta_{ms} = \frac{MC_t}{SR_t} \tag{11.2}$$

where

θ_{ms} is the index parameter.
MC_t is the total maintenance cost.
SR_t is the total sales revenue.

According to some documents, the average expenditure of the maintenance activity for all industry is about 5% of sales revenue. However, there is a quite wide variation among industries. For example, the expenditure for the maintenance activity for steel and chemical industries is approximately 12.8% and 6.8% of sales revenue, respectively.

11.4.3 INDEX III

This index relates the total maintenance cost to the total investment in plant and equipment and is expressed by

$$\theta_{mi} = \frac{MC_t}{IPE_t} \qquad (11.3)$$

where

θ_{mi} is the index parameter.
MC_t is the total maintenance cost.
IPE_t is the total amount of investment in plant and equipment.

The approximate average values for this index in the steel and chemical industries are 8.6% and 3.8%, respectively.

11.4.4 INDEX IV

This index relates the total maintenance cost to the total output (i.e., in units such as tons and megawatts) by the organization in question and is expressed by

$$\theta_{mo} = \frac{MC_t}{TO} \qquad (11.4)$$

where

θ_{mo} is the index parameter.
MC_t is the total maintenance cost.
TO is the total output by the organization in question, expressed in units such as tons, megawatts, and gallons.

11.4.5 INDEX V

This index is frequently used in material control areas and is defined by

$$\theta_{mc} = \frac{PJAM_t}{PJ_t} \qquad (11.5)$$

where

θ_{mc} is the index parameter.
$PJAM_t$ is the total number of planned jobs awaiting material.
PJ_t is the total number of planned jobs.

11.4.6   INDEX VI

This index measures the maintenance budget plan accuracy and is expressed by

$$\theta_{ba} = \frac{AMC}{BMC} \tag{11.6}$$

where

θ_{ba} is the index parameter.
AMC is the actual maintenance cost.
BMC is the budgeted maintenance cost.

It is to be noted that large variances in the values of this index indicate the need for immediate attention.

11.4.7   INDEX VII

This index relates the total maintenance cost to the total manufacturing cost and is defined by

$$\theta_{mm} = \frac{MC_t}{TMC} \tag{11.7}$$

where

θ_{mm} is the index parameter.
MC_t is the total maintenance cost.
TMC is the total manufacturing cost.

11.4.8   INDEX VIII

This index relates maintenance cost to total man-hours worked and is expressed by

$$\theta_{mm} = \frac{MC_t}{MHW_t} \tag{11.8}$$

where

θ_{mm} is the index parameter.
MC_t is the total maintenance cost.
MHW_t is the total number of man-hours worked.

11.4.9   INDEX IX

This index is used for measuring maintenance effectiveness in regard to man-hours associated with emergency and unscheduled jobs and total maintenance man-hours worked. The index is defined by

$$\theta_{mh} = \frac{TMH_{eu}}{MHW_t} \tag{11.9}$$

where

θ_{mh} is the index parameter.
TMH_{eu} is the total number of man-hours associated with emergency and unscheduled jobs.
MHW_t is the total number of maintenance man-hours worked.

11.5   SAFETY IN MAINTENANCE

Nowadays, safety in maintenance is becoming a very important issue, as accidents occurring during maintenance work or concerning maintenance are increasing significantly. For example, in 1994, approximately 13.61% of all accidents in the U.S. mining industrial sector occurred during maintenance work and they have been increasing at a quite significant rate annually since 1990 [14, 15].

Some of the main reasons for safety-related problems in maintenance are poorly written maintenance instructions and procedures, poor equipment design, inadequate work tools, poor management, poor safety-related standards and tools, poor training of maintenance personnel, poor work environments, and insufficient time to conduct required maintenance-related tasks [15].

One of the important ways for improving maintenance safety is to reduce the requirement for maintenance as much as possible in systems/products during their design phase. However, when the need for maintenance cannot be avoided, designers should follow guidelines such those presented below for improving safety in maintenance [16]:

- Develop the design in a manner that reduces the probability of maintenance personnel being injured by electric shock, escaping high-pressure gas, contact with a hot surface, etc.
- Eradicate the need for conducting maintenance and adjustments close to hazardous operating parts or equipment.

- Incorporate appropriate devices or other measures for early prediction and detection of all potential failures of that the required maintenance can be conducted prior to failure with somewhat reduced risk of hazard.
- Provide appropriate guards against moving parts or articles and interlocks for blocking access to hazardous locations.
- Design for easy accessibility so that items requiring maintenance can easily be serviced, checked, removed, or replaced.
- Keep design as simple as possible, because complexity generally adds to maintenance problems.
- Incorporate appropriate fail-safe designs to prevent injury and damage if a failure occurs.
- Develop designs or procedures that minimize the occurrence of maintenance errors.
- Aim to eliminate the requirement for special tools or equipment.

PROBLEMS

1. Discuss at least four facts and figures concerning engineering maintenance.
2. Discuss the need for maintenance.
3. List at least eight important objectives of engineering maintenance.
4. Define the index that relates the total maintenance cost to the total investment in plant and equipment.
5. What are the main reasons of safety-related problems in engineering maintenance?
6. List at least six useful guidelines for equipment designers for improving safety in the maintenance activity?
7. In your opinion, what is the most important maintenance index or measure?
8. Define the frequently used index in material control areas.
9. Define the index considered quite useful for measuring inspection effectiveness.
10. Define at least three measures or indexes that can be used in the area of engineering maintenance.

REFERENCES

1. Zweekhorst, A., Evaluation of Maintenance, *Maintenance Technology*, October 1996, pp. 9–14.
2. Tesdahl, S.A., Tomlingson, P.D., Equipment Management Breakthrough Maintenance Strategy for the 21st Century, *Proceedings of the First International Conference on Information Technologies in the Minerals Industry*, December 1997, pp. 39–58.
3. Latino, C.J., *Hidden Treasure: Eliminating Chronic Failures Can Cut Maintenance Costs up to 60%*, Reliability Center, Hopewell, VA, 1999.
4. *1997 DOD Budget: Potential Reductions to Operation and Maintenance Program*, U.S. General Accounting Office, Washington, DC, 1996.
5. *Report on Infrastructure and Logistics*, Department of Defense, Washington, DC, 1995.
6. Niebel, B.W., *Engineering Maintenance Management*, Marcel Dekker, New York, 1994.
7. *Report by the Working Party on Maintenance Engineering*, Department of Industry, London, 1970.

8. Kelly, A., *Management of Industrial Maintenance*, Newes-Butterworths, London, 1978.
9. Kumar, V.D., New Trend in Aircraft Reliability and Maintenance Measures, Journal of Quality in Maintenance Engineering, Vol. 5, No. 4, 1999, pp. 287–299.
10. AMCP 706-132, *Engineering Design Handbook: Maintenance Engineering Techniques*, Department of the Army, Washington, DC, 1975.
11. Hartmann, E., Knapp, D.J., Johnstone, J.J., Ward, K.G., *How to Manage Maintenance*, American Management Association, New York, 1994.
12. Stoneham, D., *The Maintenance Management and Technology Handbook*, Elsevier Science, Oxford, U.K., 1998.
13. Westerkemp, T.A., *Maintenance Manager's Standard manual*, Prentice Hall, Paramus, NJ, 1997.
14. Accidents Facts, National Safety Council, Chicago, IL, 1999.
15. Dhillon, B.S., *Engineering Maintenance: A Modern Approach*, CRC Press, Boca Raton, Florida, 2002.
16. Hammer, W., *Product Safety Management and Engineering*, Prentice Hall, Englewood Cliffs, New Jersey, 1980.

12 Maintenance Management and Control

12.1 INTRODUCTION

The management and control of maintenance-related activities are equally important to performing maintenance. Maintenance management may be described as the function of providing policy guidance for maintenance-related activities, in addition to exercising technical and management control of maintenance programs [1, 2]. Generally, as the maintenance activity's size and group increases, the need for better management and control becomes essential.

In the past, the typical size of a maintenance group in a manufacturing establishment varied from around 5% to 10% of the operating group [3]. Today, the proportional size of the maintenance effort compared to the operating group has increased significantly, and this increase is expected to continue. The main factor behind this trend is the tendency in the industrial sector to increase the mechanization and automation of many processes. Consequently, this means lesser need for operators but higher requirement for maintenance personnel.

This chapter presents various important aspects of maintenance management and control.

12.2 MAINTENANCE MANAGEMENT-RELATED PRINCIPLES

Over the years, many maintenance management-related principles have been developed. Six important maintenance management-related principles are as follows [4, 5]:

- **Principle I: Job control depends on definite, individual responsibility for each task during a work order's life span.** A maintenance department's responsibility is to develop, implement, and provide operating support for the planning and scheduling of maintenance work. More clearly, it is the responsibility of management personnel for ensuring effective and complete use of the system within their sphere of control.
- **Principle II: Measurement comes before control.** When an individual is assigned a task to be conducted using an effective method in a specified period of time, he or she becomes automatically aware of management expectations. Control starts when management personnel compare the results against set goals.
- **Principle III: The customer service relationship is the basis of an effective maintenance organization.** Good maintenance service is very important for effectively maintaining facilities at an expected level. The team approach fostered by the organizational structure is quite important to consistent, active control of maintenance function.

DOI: 10.1201/9781003365204-12

- **Principle IV: The optimal size of a crew is the minimum number that can conduct a given task in an effective manner.** Past experiences over the years clearly indicate that most tasks require just one person.
- **Principle V: Maximum productivity occurs when each employee in an organization has a defined task to conduct in a definitive fashion and a definite time.** This principle was formulated by Frederick Taylor in the late nineteenth century, and it is still a quite important factor in management.
- **Principle VI: Schedule all control points effectively.** Schedule appropriate control points at intervals so that all the problems are detected in time and the job's scheduled completion is not delayed.

12.3   MAINTENANCE DEPARTMENT ORGANIZATION AND FUNCTIONS

There are many factors that determine the place of maintenance in the plant organization, including complexity, size, and product produced. The four guidelines considered quite useful in planning a maintenance organization are as follows [6]:

- Establish reasonably clear division of authority with minimal overlap.
- Optimize the number of persons reporting to an individual.
- Fit the organization to the personalities involved.
- Keep vertical lines of authority and responsibility as short as possible.

One of the first considerations in planning a maintenance organization is to decide whether it is beneficial to have a centralized or decentralized maintenance function. Normally, centralized maintenance serves quite well in small- and medium-sized enterprises housed in one structure, or service buildings located in an immediate geographic area. Some of the benefits and drawbacks of centralized maintenance are as follows [3]:

Benefits
- Greater use of special equipment and specialized maintenance persons.
- Normally allows more effective on-the-job training.
- More efficient compared to decentralized maintenance.
- More effective line supervision.
- Permits procurement of more modern facilities.
- Fewer maintenance personnel required.

Drawbacks
- More difficult supervision because of remoteness of maintenance site from the centralized headquarters.
- Require more time getting to and from the work area or job.
- Greater transportation cost due to remote maintenance work.
- No one person becomes totally familiar with complex hardware or equipment.

In the case of decentralized maintenance, a maintenance group is assigned to a particular area or unit. Some of the important reasons for the decentralized maintenance are to lower travel time to and from maintenance jobs, generally closer supervision, greater chances for maintenance workers to become familiar with sophisticated equipment or facilities, and a spirit of cooperation between production and maintenance workers.

Past experiences over the years clearly indicate that in large plants, a combination of centralized and decentralized maintenance generally works best. The main reason is that the benefits of both the systems can be achieved with essentially a low number of drawbacks. Nonetheless, no one specific type of maintenance organization is useful for all types of enterprises.

A maintenance department is expected to conduct a wide range of functions including the ones presented below [6–8].

- Conducting preventive maintenance (PM); more specifically, developing and implementing a regularly scheduled work program for the purpose of maintaining satisfactory equipment/facility operation as well as preventing major problems.
- Managing inventory for ensuring that parts/materials necessary to conduct maintenance tasks are readily available.
- Developing effective techniques for keeping operation personnel, upper-level management, and other concerned groups aware of maintenance activities.
- Developing contract specifications and inspecting work performed by contractors for ensuring compliance with contractual requirements.
- Planning and repairing equipment/facilities to acceptable standards.
- Training maintenance staff and other concerned personnel to improve their skills and perform effectively.
- Preparing realistic budgets that detail maintenance personnel and material needs.
- Implementing methods for improving workplace safety and developing safety education-related programs for maintenance staff.
- Keeping records on services, equipment, etc.
- Developing effective approaches for monitoring the activities of maintenance staff.
- Reviewing plans for new facilities, installation of new equipment, etc.

12.4   EFFECTIVE MAINTENANCE MANAGEMENT ELEMENTS

There are many elements of effective maintenance management whose effectiveness is the key for the overall success of the maintenance activity. Eight of these elements are described in the following eight subsections [4, 8]:

12.4.1   MAINTENANCE POLICY

A maintenance policy is one of the most important elements of effective maintenance management. It is absolutely essential for continuity of operations and a clear understanding of the maintenance management program, irrespective of the size of a maintenance organization. Generally, maintenance organizations have manuals containing items such as objectives, policies, programs, responsibilities, and authorities for all levels of supervision, useful methods and techniques, reporting-related requirements, and performance measurement indices. Lacking such documentation, that is, a policy manual, a policy document must be developed containing all essential policy-related information.

12.4.2 Job Planning and Scheduling

Job planning is a very important element of the effective maintenance management. A number of tasks may have to be conducted prior to commencement of a maintenance job; for example, procurement of tools, parts, and materials, coordination and delivery of tools, parts, and materials, identification of methods and sequencing, coordination with other departments, and securing safety-related permits.

Although the degree of planning required may vary with the craft involved and the methods used, past experience over the years clearly indicates that on average, one planner is needed for every 20 craft persons. Strictly speaking, formal planning should cover 100% of the maintenance workload, but emergency jobs and small, straightforward work assignments are carried out in a less formal environment. Thus, in most maintenance organizations, 80–85% planning coverage is attainable.

Maintenance scheduling is as important as job planning. Schedule effectiveness is based on the reliability of the planning function. For large jobs, in particular those requiring multi-craft coordination, serious consideration must be given to using methods such as Program Evaluation and Review Technique (PERT) and Critical Path Method (CPM) for assuring effective overall control. The CPM method is described in detail later in this chapter.

12.4.3 Equipment Records

Equipment records play a very important role in effectiveness and efficiency of the maintenance organization. Normally, equipment records are grouped under the following four classifications:

- **Classification I: Maintenance work performed.** This classification contains chronological documentation of all repairs and PM performed during the item's service life to date.
- **Classification II: Maintenance cost.** This classification contains historical profiles and accumulations of labor and material costs by item.
- **Classification III: Inventory.** Usually, information on inventory is provided by the stores or accounting department. Nonetheless, this classification contains information such as property number, size and type, date manufactured or acquired, procurement cost, manufacturer, and location of the item/equipment.
- **Classification IV: Files.** This classification contains information such as operating and service manuals, warranties, and drawings. Equipment records are useful when procuring new equipment/items for determining

operating performance trends, investigating incidents, troubleshooting breakdowns, highlighting areas of concern, making replacement or modification decisions, conducting life cycle cost and design studies, and performing reliability and maintainability studies.

12.4.4   MATERIAL CONTROL

Past experience over the years clearly indicates that, on average, material costs account for around 30-40% of total direct maintenance costs [4]. Efficient utilization of personnel depends largely on effectiveness in material coordination. Material-related problems can lead to false starts, unmet due dates, delays, excessive travel time, etc. Steps such as job planning, coordinating with purchasing, coordinating with stores, coordination of issuance of materials, and reviewing the completed job can help reduce material-associated problems.

Finally, it is added that deciding whether to keep spares in storage is one the most important problems of material control.

12.4.5   BACKLOG CONTROL AND PRIORITY SYSTEM

The amount of backlog within a maintenance organization is one of the determining factors of maintenance management effectiveness. Identification of backlogs is important for balancing manpower and workload requirements. Furthermore, decisions concerning overtime, subcontracting, shop assignments, hiring, etc., are largely based on backlog-related information. Management makes use of various indices for making backlog-associated decisions.

The determination of job priority in a maintenance organization is necessary since it is not possible for starting every job the day it is requested. In assigning job priorities, it is very important to consider factors such as importance of the system/item, required due dates, the length of time the job awaiting scheduling will take place, and the type of maintenance.

12.4.6   WORK ORDER SYSTEM

A work order authorizes and directs an individual or a group for performing a given task. A well-defined work order system should cover all the maintenance jobs requested and accomplished, whether one-time or repetitive jobs. The work order system is quite useful for management in controlling costs and evaluating job performance.

Although the type and size of the work order can vary significantly from one maintenance organization to another, a work order should at least contain information such as requested and planned completion dates, planned start date, work description and its reasons, item/items to be affected, material and labor costs, work category (i.e., repair, PM, installation, etc.), and appropriate approval signatures.

12.4.7   PREVENTIVE AND CORRECTIVE MAINTENANCE

The basic purpose of conducting PM is to keep equipment/facility in satisfactory condition through inspection and correction of early-stage deficiencies. Three principal factors that shape the requirement and scope of the PM effort are process reliability, economics, and standards compliance.

A major proportion of a maintenance organization's effort is spent on corrective maintenance (CM). Thus, CM is a very important factor in the effectiveness of maintenance organization.

12.4.8 PERFORMANCE MEASUREMENT

Successful maintenance organizations regularly measure their performance through various means. Performance analyses contribute to maintenance department efficiency and are essential to revealing the equipment downtime, developing plans for future maintenance, peculiarities in operational behavior of the concerned organization, etc. Various types of performance-related indices for use by the maintenance management are presented later in this chapter.

12.5 QUESTIONS FOR EVALUATING MAINTENANCE PROGRAM EFFECTIVENESS

The U.S. Energy Research and Development Administration performed a study on maintenance management-associated matters and formulated the following ten question for maintenance managers to self-evaluate their ongoing maintenance effort [4]:

1. Do you ensure that maintainability factors are considered appropriately during the design of new or modified equipment/facilities?
2. Are you providing the craft persons with correct quality and quantity of material when and where they need it?
3. Are you aware of what equipment/facility and activity consume most of the maintenance money?
4. Do you have an effective base to conduct productivity measurements, and is productivity improving?
5. Have you balanced your spare parts inventory in regard to carrying cost vs. anticipated downtime losses?
6. Are you aware if the craft persons use proper tools to conduct their tasks?
7. With respect to job-related costs, are you in a position to compare with the "should" with the "what"?
8. Are you aware of how your craft persons spend their time, that is, travel, delays, etc.?
9. Are you aware of how much time your foreman spends at the desk and at the job site?
10. Are you aware of whether safety-related practices are being followed?

If an unqualified "yes" is the answer to each of the above ten questions, then your maintenance program is on a sound footing to meet the organizational objectives. Otherwise, appropriate corrective measures are needed.

12.6   MAINTENANCE PROJECT CONTROL METHODS

Two widely used maintenance project control methods are Program Evaluation and Review Technique (PERT) and Critical Path Method (CPM). The development of PERT is associated with the U.S. Polaris project for monitoring the effort of 250 prime contractors and 9000 subcontractors. PERT was the result of efforts of a team formed by the U.S. Navy's Special Office in 1958. The members of the team included the consulting firm of Booz, Allen, and Hamilton and the Lockheed Missile System Division [9–12].

The history of CPM goes back to 1956 when E.I. DuPont de Nemours and Co. used a network model to schedule design and construction-related activities. The following year (i.e., 1957), CPM was used in the construction of a $10 million chemical plant in Louisville, Kentucky.

In maintenance and other projects, three important factors of concern are resource availability, cost, and time. CPM and PERT deal with these factors in combination and individually.

PERT and CPM are quite similar. The major difference between the two is that when the completion times of activities of the project are uncertain, PERT is used and with the certainty of completion times, CPM is utilized [11].

The following six steps are involved with PERT and CPM [9]:

- **Step I:** Break a project into individual tasks or jobs.
- **Step II:** Arrange these tasks/jobs into a logical network.
- **Step III:** Determine duration time of each task/job.
- **Step IV:** Develop a schedule.
- **Step V:** Highlight tasks/jobs that control the completion of project.
- **Step VI:** Redistribute resources or funds for improving schedule.

The following two subsections present a formula for estimating activity expected duration times and CPM in detail.

12.6.1   ACTIVITY EXPECTED DURATION TIME ESTIMATION

The PERT scheme calls for three estimates of activity duration time using the following formula for calculating the final time:

$$T_{ae} = \frac{\alpha + 4\theta + \beta}{6} \tag{12.1}$$

where

T_{ae} is the activity expected duration time.
α is the optimistic or minimum time an activity will require for completion.
β is the pessimistic or maximum time an activity will require for completion.
θ is the most likely time an activity will require for completion. It is to be noted that this is the time also used for CMP activities.

It is to be noted that Equation (12.1) is based on Beta distribution [13].

Example 12.1

Assume that we have the following time estimates to accomplish an activity:

$\alpha = 40\,\text{days}$

$\beta = 90\,\text{days}$

$\theta = 65\,\text{days}$

Calculate the activity expected duration time.
   By substituting the given data values into Equation (12.1), we obtain

$$T_{ae} = \frac{40 + 4(65) + 90}{6} = 65\,\text{days}$$

Thus, the expected duration time for the activity is 65 days.

12.6.2 Critical Path Method (CPM)

Four symbols used to construct a CPM network are shown in Figure 12.1.

The circle in Figure 12.1(i) denotes an event. More clearly, it represents an unambiguous point in the life of a project. An event could be the start or completion of an activity or activities, and generally, the events are labeled by number. The circle shown in Figure 12.1(ii) with three divisions is also denotes an event. Its top half labels the event with a number, and the bottom portions indicate earliest event time (EET) and latest event time (LET). EET is the earliest time in which an activity can be accomplished or an event could be reached. Similarly, LET is the latest time in which an event can be reached without delaying project completion.

The continuous arrow in Figure 12.1(iii) represents an activity that consumes time, money, and manpower. This arrow always starts at a circle and ends at a circle. Finally, the dotted arrow in Figure 12.1(iv) denotes a dummy activity or a restraint. More clearly, this is an imaginary activity that does not consume time, money, or manpower. An application of a dummy activity is shown in Figure 12.2. It shows that activities A and B must be accomplished before activity C can start. However, only activity B must be completed prior to starting activity D.

Example 12.2

Assume that a maintenance project was broken down into a set of six activities, after which Table 12.1 was prepared. Prepare a CPM network using Figure 12.1 symbols and Table 12.1 data values, and determine the critical path associated with the network.

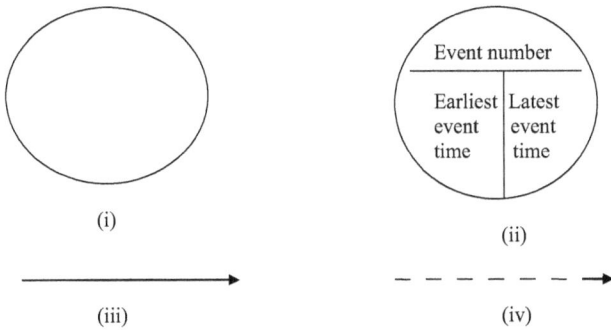

FIGURE 12.1 Critical path method symbols: (i) circle, (ii) circle with divisions, (iii) continuous arrow, (iv) dotted arrow.

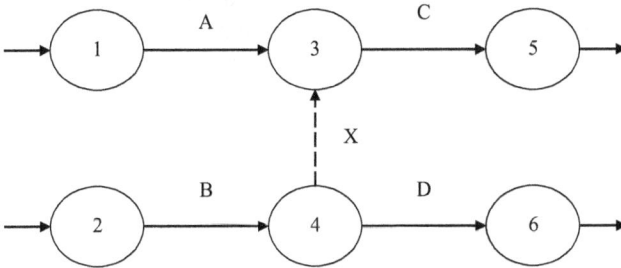

FIGURE 12.2 A portion of a CPM network with a dummy activity X.

TABLE 12.1
Maintenance Project Activities' Associated Data Values

Activity No.	Activity Identification	Immediate Predecessor Activity or Activities	Expected Duration in Days
1	A	—	10
2	B	—	2
3	C	A,B	2
4	D	A	5
5	E	D	3
6	F	C,E	8

A CPM network for given data values in Table 12.1 is presented in Figure 12.3. In this figure, the following paths originate and terminate at events 1 and 6, respectively:

- B-C-F (2 + 2 + 8 = 12 days)
- A-Y-C-F (10 + 0 + 2 + 8 = 20 days)
- A-D-E-F (10 + 5 + 3 + 8 = 26 days)

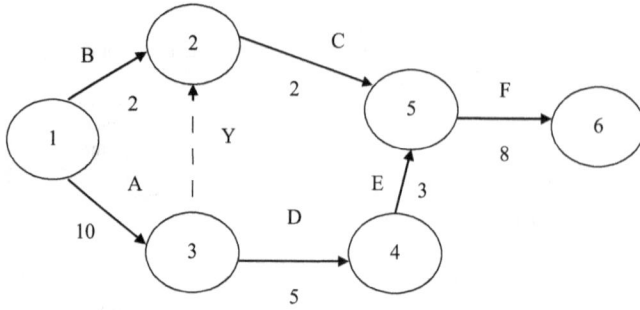

FIGURE 12.3 A CPM network for Table 12.1 data values.

The quantities in parentheses above show the total time in days for each path. The dummy activity consumes zero time and by definition, the longest path through the network is the critical path. Inspection of the above three values show that 26 days is the largest time. More specifically, it will take 26 days from event 1 to reach event 6. Thus, this is the critical path. The word "critical" is used because any delay in the completion of activities along this path (i.e., critical path) will result in delay of maintenance project's completion.

12.6.3 CRITICAL PATH DETERMINATION APPROACH

It is to be noted that for simple and straightforward CPM networks, the critical path can easily be identified in a manner discussed above. However, for complex networks, a more systematic approach is needed. This section presents one such approach with the aid of Figure 12.4.

The symbols used in Figure 12.4 are defined as follows:

- $EET(i)$ is the earliest event time of event i.
- $LET(i)$ is the latest event time of event i.
- $EET(j)$ is the earliest event time of event j.
- $LET(j)$ is the latest event time of event j.
- $D(i,j)$ is the expected duration time of the activity between events i and j.

The following five steps are associated with the approach:

- **Step I:** Construct CPM network.
- **Step II:** Calculate EET of each event by making a forward pass of the network and using: For any event j,

$EET(j)$ = Maximum for all preceding i of

$$\left[EET(i) + D(i,j) \right]$$

(12.2)

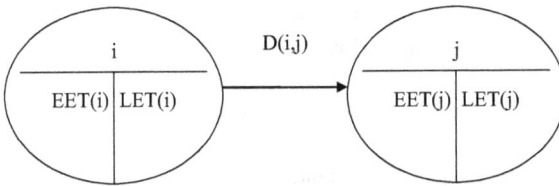

FIGURE 12.4   A single activity CPM network.

Also,

$$EET\left(\text{first event}\right) = 0 \tag{12.3}$$

- **Step III:** Calculate *LET* of each event by making a backward pass of the network and using: For any event i,

$LET(i)$ = Maximum of all succeeding

$$j\,\text{of}\left[LET\left(j\right) + D\left(i, j\right)\right] \tag{12.4}$$

Also,

$$LET\left(\text{last event}\right) = EET\left(\text{last event}\right) \tag{12.5}$$

If *LET* of all events of the network in question was calculated correctly, we should get

$$LET\left(\text{first event}\right) = 0 \tag{12.6}$$

- **Step IV:** Select network events with equal *EET* and *LET*. If the network results in only one path, that is, from the first event to the last event, with *EET* = *LET*, this path is critical. Otherwise, go to the next step.
- **Step V:** Calculate the total float for each activity on each of the paths with *EET* = *LET*. The critical path is the path that results in the least sum of the total floats. The total float for any activity (i,j) can be calculated by using the following equation:

$$\text{Total float} = LET\left(j\right) - EET\left(i\right) - D\left(i, j\right) \tag{12.7}$$

Additional information on this approach is available in Ref. [14].

12.6.4   CPM BENEFITS AND DRAWBACKS

As with other methods, CPM has its benefits and drawbacks. Some of the CPM benefits are as follows [15]:

- It highlights activities important for completing the project on time. These activities must be completed on time to accomplish the entire project on time.
- It helps avoiding duplications and omissions as well as determining project duration systematically.
- It helps improve project understanding and communication among involved personnel.
- It shows interrelationships in workflow and is quite useful in determining labor and resources needs in advance.
- It is a quite effective tool for controlling costs and can easily be computerized.
- It is a quite effective tool for monitoring project progress.

Some of the drawbacks of the CPM are as follows [15]:

- Time-consuming.
- Inclination for using pessimistic estimates for activity times.
- Costly.
- Poor estimates of activity times.

12.7 MAINTENANCE MANAGEMENT CONTROL INDICES

Management uses various approaches for measuring effectiveness of the maintenance function. Quite often, it uses indices to manage and control maintenance. These indices show trends by using past data as a reference point. Generally, a maintenance organization uses various indices for measuring maintenance effectiveness, as there is no single index that can accurately reflect the overall performance of the maintenance activity. The main objective of these indices is to encourage maintenance management for improving on past performance.

This section presents 15 broad and specific indices [3, 5, 16, 17]. The broad indices indicate the organization's overall performance with respect to maintenance and the specific indices indicate the performance in particular areas of the maintenance function. The values of all these indices are plotted periodically to show trends.

Both broad and specific indicators (i.e., indices) are presented in the following two subsections:

12.7.1 BROAD INDICATORS

This subsection presents three such indicators.

12.7.1.1 Index I
This is defined by

$$\theta_1 = \frac{C_{tm}}{O_t}$$

(12.8)

where

θ_1 is the index parameter.
C_{tm} is the total maintenance cost.
O_t is the total output expressed in tons, gallons, megawatts, etc.

This index relates the total maintenance cost to the total output by the organization.

12.7.1.2   Index II
This is expressed by

$$\theta_2 = \frac{C_{tm}}{I_t}$$ (12.9)

where

θ_2 is the index parameter.
C_{tm} is the total maintenance cost.
I_t is the total investment in plant and equipment.

This index relates the total maintenance cost to the total investment in plant and equipment. The approximate average figures for θ_2 in the chemical and steel industries are 3.8% and 8.6%, respectively.

12.7.1.3   Index III
This is defined as follows:

$$\theta_3 = \frac{C_{tm}}{S_t}$$ (12.10)

where

θ_3 is the index parameter.
C_{tm} is the total maintenance cost.
S_t is the total sales.

Past experience over the years indicates that average expenditure for maintenance for all industries was about 5% of sales. However, there was a quite wide variation among industries. For example, the average values of θ_3 for chemical and steel industries were 6.8% and 12.8%, respectively.

12.7.2   SPECIFIC INDICATORS

This subsection presents 12 such indicators.

12.7.2.1 Index IV

This index can be used to measures the accuracy of the maintenance budget plan and is defined by

$$\theta_4 = \frac{C_{am}}{C_{bm}}$$ (12.11)

where

θ_4 is the index parameter.
C_{am} is the total actual maintenance cost.
C_{bm} is the total budgeted maintenance cost.

In this case, large variances indicate the need for immediate attention.

12.7.2.2 Index V

This index is useful in scheduling work and is defined by

$$\theta_5 = \frac{J_{pc}}{J_p}$$ (12.12)

where

θ_5 is the index parameter.
J_{pc} is the total number of planned jobs completed by established due dates.
J_p is the total number of planned jobs.

The value of θ_5 should be high for keeping backlogs down.

12.7.2.3 Index VI

This index can be used to measure maintenance effectiveness and is expressed as follows:

$$\theta_6 = \frac{MH_{eu}}{MH_m}$$ (12.13)

where

θ_6 is the index parameter.
MH_{eu} is the man-hours of emergency and unscheduled jobs.
MH_m is the total maintenance man-hours worked.

12.7.2.4 Index VII

This is a useful index to monitor progress in cost reduction efforts and is expressed by

$$\theta_7 = \frac{P_{mh}}{C_{mp}}$$ (12.14)

where

θ_7 is the index parameter.
P_{mh} is the percentage of maintenance man-hours spent on scheduled jobs.
C_{mp} is the maintenance cost per unit of production.

12.7.2.5 Index VIII

This index is useful in material control area and is expressed by

$$\theta_8 = \frac{J_{pam}}{J_p} \tag{12.15}$$

where

θ_8 is the index parameter.
J_{pam} is the total number of planned jobs awaiting material.
J_p is the total number of planned jobs.

12.7.2.6 Index IX

This index relates material and labor costs and is defined by

$$\theta_9 = \frac{C_{ml}}{C_{mm}} \tag{12.16}$$

where

θ_9 is the index parameter.
C_{ml} is the total maintenance labor cost.
C_{mm} is the total maintenance materials cost.

12.7.2.7 Index X

This index relates maintenance cost to man-hours worked and is defined by

$$\theta_{10} = \frac{C_{tm}}{H_{mw}} \tag{12.17}$$

where

θ_{10} is the index parameter.
C_{tm} is the total maintenance cost.
H_{mw} is the total number of man-hours worked.

12.7.2.8   Index XI

This index can also be used to measure maintenance effectiveness and is defined by

$$\theta_{11} = \frac{DT_b}{DT_t} \tag{12.18}$$

where

θ_{11} is the index parameter.
DT_b is the downtime caused by breakdowns.
DT_t is the total downtime.

12.7.2.9   Index XII

This is a useful index for maintenance overhead control and is defined by

$$\theta_{12} = \frac{C_{ma}}{C_{tm}} \tag{12.19}$$

where

θ_{12} is the index parameter.
C_{ma} is the total maintenance administration cost.
C_{tm} is the total maintenance cost.

12.7.2.10   Index XIII

This index relates labor and material costs and is defined by

$$\theta_{13} = \frac{C_{tm}}{C_m} \tag{12.20}$$

where

θ_{13} is the index parameter.
C_{tm} is the total maintenance cost.
C_m is the total manufacturing cost.

12.7.2.11   Index XIV

This is a very important index used to measure inspection effectiveness and is expressed by

$$\theta_{14} = \frac{J_i}{I_c} \tag{12.21}$$

where

θ_{14} is the index parameter.
J_i is the number of jobs resulting from inspections.
I_c is the total number of inspections completed.

12.7.2.12 Index XV

This is a useful index to control PM activity within a maintenance organization and is expressed by

$$\theta_{15} = \frac{T_{ppm}}{T_{emf}} \tag{12.22}$$

where

θ_{15} is the index parameter.
T_{ppm} is the total time spent in performing PM.
T_{emf} is the total time spent for the entire maintenance function.

As per the past experience, the value of θ_{15} should be kept within 20% and 40% limits.

PROBLEMS

1. Discuss six important maintenance management-related principles.
2. List at least nine important functions of a maintenance department.
3. What are the benefits and drawbacks of centralized maintenance?
4. Discuss the following two elements of effective maintenance management:
   - Job planning and scheduling
   - Equipment records
5. List at least seven questions the maintenance managers can use to self-evaluate effectiveness of their overall maintenance management.
6. Discuss the history of PERT and CPM. What are the six steps involved with PERT and CPM?
7. Assume that we have the following time estimates to accomplish an activity the PERT scheme calls for, to estimate the final time:

   $\alpha = 30$ days
   $\beta = 80$ days
   $\theta = 55$ days

   where

   α is the optimistic or minimum time the activity will require for completion.
   β is the pessimistic or maximum time the activity will require for completion.
   θ is the most likely time the activity will require for completion.

   Calculate the activity expected duration time.

8. What are the advantages and disadvantages of CPM?
9. Define an index that can be used to evaluate overall performance of a maintenance organization.
10. Define an index that can be used to measure the accuracy of the maintenance budget plan.

REFERENCES

1. AMCP 706-132, *Engineering Design Handbook: Maintenance Engineering Techniques, Department of the Army*, Washington, DC, June 1975.
2. DOD Int. 4151.12, *Policies Governing Maintenance Engineering within the Department of Defense*, Washington, DC, June 1968.
3. Neibel, B.W., *Engineering Maintenance Management*, Marcel Dekker, New York, 1994.
4. ERHQ-0004, *Maintenance Manager's Guide, Energy Research and Development Administration*, Washington, DC, 1976.
5. Westerkamp, T.A., *Maintenance Manager's Standard Manual*, Prentice Hall, Paramus, New Jersey, 1997.
6. Higgins, L.R., *Maintenance Engineering Handbook*, McGraw-Hill, New York, 1988.
7. Jordon, J.K., *Maintenance Management*, American Water Works Association, Denver, Colorado, 1990.
8. Dhillon, B.S., *Engineering Management*, Technomic Publishing Company, Lancaster, Pennsylvania, 1987.
9. Malcolm, D.G., Roseboom, J.H., Clark, C.E., Fazar, W., Application of Technique for Research and Development Program Evaluation, *Operations Research*, Vol. 7, 1959, pp. 646–669.
10. Riggs, J.L., Inoue, M.S., *Introduction to Operations Research and Management Science: A General Systems Approach*, McGraw-Hill, New York, 1974.
11. Chase, R.B., Aquilano, N.J., *Production and Operations Research: A Life Cycle Approach*, Richard D. Irwin, Homewood, Illinois, 1981.
12. Lee, S.M., Moore, L.J., Taylor, B.W., *Management Science*, Wm.C. Brown Co., Duduque, Iowa, 1981.
13. Clark, C.E., The PERT Model for the Distribution of an Activity Time, *Operations Research*, Vol. 10, 1962, pp. 405–406.
14. Dhillon, B.S., *Engineering Maintenance: A Modern Approach*, CRC Press, Boca Raton, Florida, 2002.
15. Lomax, P.A., *Network Analysis: Applications to the Building Industry*, The English Universities Press Limited, London, 1969.
16. Hartmann, E., Knapp, D.J., Johnstone, J.J., Ward, K.G., *How to Manage Maintenance*, American Management Association, New York, 1994.
17. Stoneham, D., *The Maintenance Management and Technology Handbook*, Elsevier Science, Oxford, 1998.

13 Preventive and Corrective Maintenance

13.1 INTRODUCTION

Preventive maintenance is a very important element of a maintenance activity, and within a maintenance department, it generally accounts for a quite significant proportion of the overall maintenance-related activity. Preventive maintenance is the care and servicing by maintenance personnel for keeping facilities in a satisfactory operational state by providing systematic inspection, detection, and correction of all incipient failures either prior to their development into major failures or before their occurrence [1, 2]. There are many objectives of conducting preventive maintenance, including minimizing critical equipment breakdowns, improving capital equipment's productive life, improving the health and safety of maintenance personnel, and reducing production losses caused by equipment failure [3].

Corrective maintenance is the remedial action conducted because of failure of deficiencies discovered during preventive maintenance or otherwise, to repair an item to its operating state [1, 2, 4, 5]. Generally, corrective maintenance is an unplanned maintenance action that needs urgent attention that must be added, integrated with, or substituted for previously scheduled work. Corrective maintenance or repair is a very important element of overall maintenance activity.

This chapter presents various important aspects of both preventive maintenance and corrective maintenance.

13.2 PREVENTIVE MAINTENANCE ELEMENTS AND A PRINCIPLE FOR SELECTING ITEMS FOR PREVENTIVE MAINTENANCE

There are following seven elements of preventive maintenance [1, 2]:

- **Element I: Inspection.** Periodically inspecting items for determining their serviceability by comparing their physical, electrical, mechanical, and other characteristics to established standards.
- **Element II: Calibration.** Detecting and adjusting any discrepancy in the accuracy of the material or parameter being compared to the established standard value.
- **Element III: Testing.** Periodically testing to determine serviceability and detect electrical and mechanical degradation.
- **Element IV: Adjustment.** Periodically making adjustments to stated variable elements to achieve optimum performance.
- **Element V: Servicing.** Periodically lubricating, charging, cleaning, and so on, materials or items to prevent incipient failures' occurrence.

DOI: 10.1201/9781003365204-13

- **Element VI: Installation.** Periodically replacing limited-life item or items experiencing time cycle or wear degradation to maintain the stated tolerance level.
- **Element VII: Alignment.** Making changes to an item's stated variable elements to achieve optimum performance.

The formula principle presented below can be quite useful in deciding whether to implement a preventive maintenance program for an item or system [6, 7].

$$(m)(C_a)(\beta) > C_{pms} \qquad (13.1)$$

where

m is the total number of breakdowns.
C_a is the average cost per breakdown.
β is 70% of the total cost of breakdowns.
C_{pms} is the total cost of the preventive maintenance system.

13.3 STEPS FOR DEVELOPING A PREVENTIVE MAINTENANCE PROGRAM

Development of an effective preventive maintenance program needs the availability of items such as skilled personnel, test instruments and tools, management support and user cooperation, manufacturer's recommendations, past data from similar equipment, service manuals, and accurate historical records of equipment [8].

A highly effective preventive maintenance program can be developed in a short time by following the six steps presented below [9]:

- **Step I: Highlight and select the areas.** Highlight and select one or two important areas on which to concentrate the initial preventive maintenance effort. The main objective of this step is to obtain good results in areas that are highly visible.
- **Step II: Highlight the preventive maintenance-related requirements.** Define the preventive maintenance-related needs and then develop a schedule for two types of tasks: daily preventive maintenance inspections and periodic preventive maintenance assignments.
- **Step III: Determine assignment frequency.** Establish assignments' frequency and review the item or equipment conditions and records. The frequency depends on factors such as the experience of personnel familiar with the equipment or item under consideration, recommendations from engineers, and vendor recommendations.
- **Step IV: Prepare the preventive maintenance-related assignments.** Prepare the periodic and daily assignments in an effective manner and then get them approved.
- **Step V: Schedule the preventive maintenance-related assignments.** Schedule the defined preventive maintenance-related assignments on the basis of a 12-month period.

- **Step VI: Expand the preventive maintenance program as appropriate.**
  Expand the preventive maintenance program to other areas on the basis of
  experience gained from the pilot preventive maintenance projects.

13.4   PREVENTIVE MAINTENANCE MEASURES

There are many preventive maintenance-associated measures. Two of these measures
taken from the published literature are presented in the following two subsections
[1, 2, 10]:

13.4.1   MEDIAN PREVENTIVE MAINTENANCE TIME

This is an important measure of preventive maintenance. It is the equipment downtime required for performing 50% of all scheduled preventive maintenance actions
under the conditions stated for median preventive maintenance times, the median
preventive maintenance time is expressed by.

$$PMT_m = \text{anti} \log \left[\frac{\sum_{i=1}^{n} \lambda_i \log PMT_{mi}}{\sum_{i=1}^{n} \lambda_i} \right] \qquad (13.2)$$

where

PMT_m is the median preventive maintenance time.
n is the total number of data points.
λ_i is the constant failure rate of component i of the equipment for which maintainability is to be determined, adjusted for factors such as tolerance and
  interaction failures, duty cycle, and catastrophic failures that will result in
  deterioration of equipment performance to the degree that a maintenance-
  associated action will be taken for $i = 1, 2, 3, ..., n$.
PTM_{mi} is the average time required to conduct i preventive maintenance task
  for $i = 1, 2, 3, ..., n$.

13.4.2   MEAN PREVENTIVE MAINTENANCE TIME

This is another important measure of preventive maintenance. It is the average equipment downtime required for performing scheduled preventive maintenance. Mean
preventive maintenance time is defined by

$$MPMT = \left[\sum_{i=1}^{n} f_i MPMT_i \right] / \sum_{i=1}^{n} f_i \qquad (13.3)$$

where

MPMT is the mean preventive maintenance time.

n is the total number of data points.

f_i is the frequency of i preventive maintenance task in tasks per operating hour after adjustment for item or equipment duty cycle.

13.5   PREVENTIVE MAINTENANCE MATHEMATICAL MODELS

Over the years, many mathematical models have been developed for performing various types of preventive maintenance. Two such models are presented in the following two subsections [2, 11, 12]:

13.5.1   MODEL I

This mathematical model represents a system that can either undergo periodic preventive maintenance or fail completely. The failed system is repaired. The state space diagram of the system is shown in Figure 13.1 [13].

This model can predict items such as probability of the system being down for preventive maintenance, probability of system failure, and system availability. The model is subject to the following two assumptions:

- **Assumption I:** System preventive maintenance, failure, and repair rates are constant.
- **Assumption II:** After preventive maintenance or repair the system is as food as new.

The symbols used to develop equations for the model are defined below:

- λ is the system failure rate.
- λ_p is the rate of the system being down for preventive maintenance.
- μ is the system repair or corrective maintenance rate.
- μ_p is the rate of system preventive maintenance performance.

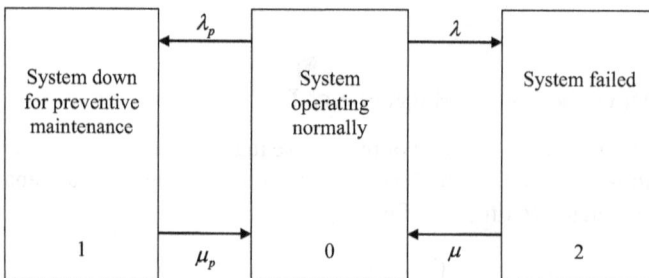

FIGURE 13.1   System state space diagram.

- j is the jth system state; $j = 0$ (system operating normally), $j = 1$ (system down for preventive maintenance), $j = 2$ (system failed).
- $P_j(t)$ is the probability that the system is in state j at time t, for $j = 0, 1, 2$.

Using the Markov method presented in Chapter 4 and Figure 13.1, we obtain the following equations [14]:

$$\frac{dP_0(t)}{dt} + \left(\lambda + \lambda_p \right) P_0(t) = \mu_p P_1(t) + \mu P_2(t) \tag{13.4}$$

$$\frac{dP_1(t)}{dt} + \mu_p P_1(t) = \lambda_p P_0(t) \tag{13.5}$$

$$\frac{dP_2(t)}{dt} + \mu P_2(t) = \lambda P_0(t) \tag{13.6}$$

At time $t = 0$, $P_0(0) = 1$, $P_1(0) = 0$, and $P_2(0) = 0$.
By solving Equations (13.4)–(13.6), we obtain

$$P_0(t) = \frac{\mu\mu_p}{M_1 M_2} + \left[\frac{(M_1 + \mu_p)(M_1 + \mu)}{M_1(M_1 - M_2)} \right] e^{M_1 t} - \left[\frac{(M_2 + \mu_p)(M_2 + \mu)}{M_2(M_1 - M_2)} \right] e^{M_2 t} \tag{13.7}$$

$$P_1(t) = \frac{\lambda_p \mu}{M_1 M_2} + \left[\frac{\lambda_p M_1 + \lambda_p \mu}{M_1(M_1 - M_2)} \right] e^{M_1 t} - \left[\frac{(\mu + M_2)\lambda_p}{M_2(M_1 - M_2)} \right] e^{M_2 t} \tag{13.8}$$

$$P_2(t) = \frac{\lambda\mu_p}{M_1 M_2} + \left[\frac{\lambda_p M_1 + \lambda\mu_p}{M_1(M_1 - M_2)} \right] e^{M_1 t} - \left[\frac{(\mu_p + M_2)\lambda}{M_2(M_1 - M_2)} \right] e^{M_2 t} \tag{13.9}$$

where

$$M_1, M_2 = \frac{-B \pm \left[B^2 - \left(\mu_p \mu + \lambda\mu_p + \lambda_p \mu \right) \right]^{1/2}}{2} \tag{13.10}$$

$$B = \mu_p + \mu + \lambda_p + \lambda \tag{13.11}$$

$$M_1 + M_2 = -B \tag{13.12}$$

$$M_1 M_2 = \mu_p \mu + \lambda_p \mu + \lambda\mu_p \tag{13.13}$$

The probability of the system being down for preventive maintenance, corrective maintenance or repair, and operating normally is given by Equation (13.8), Equation (13.9), and Equation (13.7), respectively.

As time t becomes very large, we obtain the following steady-state equations from Equation (13.7)–Equation (13.9), respectively:

$$P_0 = \frac{\mu\mu_p}{M_1 M_2} \tag{13.14}$$

$$P_1 = \frac{\lambda_p \mu}{M_1 M_2} \tag{13.15}$$

$$P_2 = \frac{\lambda \mu_p}{M_1 M_2} \tag{13.16}$$

where

P_0, P_1, and P_2 are the steady-state probabilities of the system being states 0, 1, and 2, respectively.

Example 13.1

Assume that in Equation (13.15), we have $\lambda = 0.0002$ failures per hour, $\lambda_p = 0.0005$ per hour, $\mu = 0.0004$ repairs per hour, and $\mu_p = 0.0008$ per hour. Calculate the steady-state probability that the system is down for preventive maintenance.

By substituting the above given data values into Equation (13.15), we obtain

$$P_1 = \frac{(0.0005)(0.0004)}{(0.0008)(0.0004) + (0.0005)(0.0004) + (0.0002)(0.0008)}$$
$$= 0.2941$$

Thus, the steady-state probability that the system will be down for preventive maintenance is 0.2941.

13.5.2 Model II

Inspections are a very important component of preventive maintenance. Generally, inspections are disruptive, but they reduce the occurrence of failures. This mathematical model is concerned with obtaining the optimum number of inspections per facility per unit of time. Total facility downtime is expressed by

$$T_{fd} = \frac{(T_f b)\theta}{n} + n(T_{fi}) \tag{13.17}$$

where

T_{fd} is the total downtime per unit of time for a given facility.
n is the number of inspections per facility per unit of time.
T_{fb} is the facility downtime per failure or breakdown.
T_{fi} is the facility downtime per inspection.
θ is a constant associated with a specific facility.

By differentiating Equation (13.17) with respect to n, we obtain

$$\frac{dT_{fd}}{dn} = -\left[\frac{(T_{fb})\theta}{n^2}\right] + T_{fi} \tag{13.18}$$

By equating Equation (13.18) to zero and then rearranging it, we get

$$n^* = \left[\frac{\theta(T_{fb})}{T_{fi}}\right]^{1/2} \tag{13.19}$$

where

n^* is the optimum number of inspections per facility per unit of time.

By substituting Equation (13.19) into Equation (13.17), we get

$$T_{fd}^* = 2\left[\theta(T_{fb})(T_{fi})\right] \tag{13.20}$$

where

T_{fd}^* is the total optimal downtime per unit of time.

Example 13.2

Assume that the following data values are associated with an engineering facility:

$T_{fb} = 0.4$ month
$T_{fi} = 0.05$ month
$\theta = 2$

Calculate the optimal number of inspections per month by using Equation (13.19).
By inserting the given data values into Equation (13.19), we obtain

$$n^* = \left[\frac{2(0.4)}{0.05}\right]^{1/2} = 4 \text{ inspections per month}$$

Thus, the optimal number of inspections per month is 4.

13.6   PREVENTIVE MAINTENANCE ADVANTAGES AND DISADVANTAGES

There are many advantages of performing preventive maintenance. Twelve important advantages of performing preventive maintenance are as follows [6, 8]:

- **Advantage I:** Increase in equipment availability.
- **Advantage II:** Improved safety.
- **Advantage III:** Increase in production revenue.
- **Advantage IV:** Reduction in need for standby equipment.
- **Advantage V:** Reduction in overtime.
- **Advantage VI:** Balanced workload.
- **Advantage VII:** Useful in promoting benefit/cost optimization.
- **Advantage VIII:** Consistency in quality.
- **Advantage IX:** Standardized procedures, times, and costs.
- **Advantage X:** Stimulation in preaction instead of reaction.
- **Advantage XI:** Performed as convenient.
- **Advantage XII:** Reduction in parts inventory.

Some of the disadvantages of conducting preventive maintenance are as follows [6, 8]:

- More frequent access to equipment.
- Exposing equipment to possible damage.
- Use of more components.
- Increase in initial costs.

13.7   CORRECTIVE MAINTENANCE TYPES

Corrective maintenance may be grouped under the following five categories [1, 2, 15]:

- **Category I: Rebuild.** This is concerned with restoring an equipment/item to a standard as close as possible to its original state in regard to performance, life expectancy, and appearance. This is accomplished through actions such as examination of all parts, complete disassembly, repair or replacement of unserviceable or worn components according to original specifications and manufacturing tolerances, and reassembly and testing to original production requirements.
- **Category II: Overhaul.** This is concerned with repairing/restoring an item/equipment to its complete serviceable state meeting requirements stated in maintenance serviceability standards, using the "inspect and repair only as appropriate" method.
- **Category III: Servicing.** This type of corrective maintenance may be required because of a corrective maintenance-related action; for example, engine repair can result in requirement for crankcase refill, welding on, and so on.

- **Category IV: Salvage.** This is concerned with the disposal of nonrepairable materials and utilization of salvaged materials from items that cannot be repaired in the overhaul, rebuild, or repair programs.
- **Category V: Fail repair.** This is concerned with restoring the failed equipment/item to its operational state.

13.8 CORRECTIVE MAINTENANCE STEPS, DOWNTIME COMPONENTS, AND STRATEGIES FOR TIME REDUCTION AT SYSTEM LEVEL

Over the years, different researchers and authors have proposed different steps for conducting corrective maintenance [4, 5]. For our purpose, we assume that corrective maintenance can be conducted in the following five steps [1]:

- **Step I: Failure recognition.** Recognizing the existence of a failure.
- **Step II: Failure localization.** Localizing the failure within the system to a specific piece of equipment item.
- **Step III: Diagnosis within the equipment/item.** Diagnosis within an equipment/item to highlight specific failed part/component.
- **Step IV: Failed part repair/replacement.** Repairing/replacing failed parts/components.
- **Step V: Return system to service.** Checking out and returning the system back to service.

Corrective maintenance downtime is made up of three major components. These major components are active repair time, administrative and logistic time, and delay time. The active repair time component is made up of six subcomponents. These subcomponents are fault location time, checkout time, spare item obtainment time, preparation time, adjustment and calibration time, and fault correction time [1, 16].

In order to improve the effectiveness of corrective maintenance, it is very important to reduce corrective maintenance time. Five strategies that are considered quite useful for reducing system-level corrective maintenance time are as follows [2, 10]:

- **Strategy I: Improve fault recognition, location, and isolation.** Past experiences over the years clearly indicate that within a corrective maintenance activity, fault recognition, location, and isolation consume the most time. Factors that help to reduce corrective maintenance-related time are well-trained maintenance personnel, good maintenance procedures, unambiguous fault isolation capability, and well-designed fault indicators.
- **Strategy II: Employ redundancy.** This is concerned with designing in redundant parts or components that can be switched in during the repair of faulty parts so that the equipment or system continues to function. In this case, although the overall maintenance workload may not be reduced, the downtime of the equipment could be impacted quite significantly.

- **Strategy III: Improve accessibility.** Past experiences over the years clearly indicate that quite often a significant amount of time is spent accessing failed parts. Careful attention to accessibility during design can help to reduce the parts' accessibility time and, consequently, the corrective maintenance time.
- **Strategy IV: Consider human factors.** During design, paying careful attention to human factors such as size, shape, and weight of components; information processing aids; selection and placement of indicators and dials; readability of instructions; and size and placement of access and gates can help lower corrective maintenance time quite significantly.
- **Strategy V: Improve interchangeability.** Effective functional and physical interchangeability is a very important factor in removing and replacing components/parts, thus lowering corrective maintenance time.

13.9　CORRECTIVE MAINTENANCE-RELATED MEASURES

There are many corrective maintenance-associated measures. Two of these measures are presented in the following two subsections [1, 10, 17]:

13.9.1　MEDIAN ACTIVE CORRECTIVE MAINTENANCE TIME

This is an important measure of corrective maintenance. Median active corrective maintenance time is a measure of the time within which 50% of all corrective maintenance activities can be conducted. It is to be noted that the computation of this measure is subject to the probability distribution describing corrective maintenance times. Thus, the median of corrective maintenance times following a lognormal distribution is defined by [2, 10].

$$CMT_{ma} = \text{anti log}\left[\frac{\sum \lambda_j \log CMT_j}{\sum \lambda_j}\right] \tag{13.21}$$

where

CMT_{ma} is the median active corrective maintenance time.
CMT_j is the corrective maintenance time of the jth equipment element.
λ_j is the failure rate of the jth equipment element.

Generally, corrective maintenance times are described by lognormal, exponential, and normal probability distributions. Examples of the type of equipment that follow these distributions are as follows:

- **Lognormal distribution:** Corrective maintenance times of electronic equipment that does not possess built-in test capability generally follow this distribution.

- **Exponential distribution:** Corrective maintenance times of electronic equipment with a good built-in test capability and rapid remove and replace maintenance concept quite often follow exponential distribution.
- **Normal distribution:** Corrective maintenance times of mechanical or electromechanical equipment with a remove and replacement concept quite often follow normal distribution.

13.9.2 MEAN CORRECTIVE MAINTENANCE TIME

This is another important measure of corrective maintenance and is expressed by

$$CMT_m = \frac{\sum \lambda_j CMT_j}{\sum \lambda_j} \qquad (13.22)$$

where

CMT_m is the mean corrective maintenance time.
λ_j is the failure rate of the jth equipment element.
CMT_j is the corrective maintenance of the jth equipment element.

13.10   CORRECTIVE MAINTENANCE MATHEMATICAL MODELS

There are many mathematical models available in the published literature that can be used in conducting corrective maintenance. This section presents two such models. These two models take into consideration item failure and corrective maintenance rates and can be used for predicting item/equipment/system availability, reliability, probability of being in a failed state (i.e., undergoing repair or corrective maintenance), mean time to failure, etc.

13.10.1   MATHEMATICAL MODEL I

This mathematical model represents a system that can be in either operating or failed state. The failed system is repaired back to its operating state. Most industrial systems/equipment/items follow this pattern. The system-state space diagram is shown in Figure 13.2. The numerals in boxes in Figure 13.2 denote system states.

The following assumptions are associated with the model:

- System failure and repair (i.e., corrective maintenance) rates are constant.
- The repaired system is as good as new.
- All system failures are statistically independent.

The symbols used for developing equations for the model are defined below:

- k is the kth system state; $k = 0$ (system operating normally), $k = 1$ (system failed).
- λ_s is the system failure rate.

- μ_{sc} is the system corrective maintenance or repair rate.
- $P_k(t)$ is the probability that the system is in state k at time t, for $k = 0$, and $k = 1$.

By using the Markov method presented in Chapter 4 and Figure 13.2, we write down the following two equations [2, 18]:

$$\frac{dP_0(t)}{dt} + \lambda_s P_0(t) = \mu_{sc} P_1(t) \tag{13.23}$$

$$\frac{dP_1(t)}{dt} + \mu_{sc} P_1(t) = \lambda_s P_0(t) \tag{13.24}$$

At time $t = 0$, $P_0(0) = 1$ and $P_1(0) = 0$.
Solving Equation (13.23) and Equation (13.24), we obtain

$$P_0(t) = \frac{\mu_{sc}}{(\lambda_s + \mu_{sc})} + \frac{\lambda_s}{(\lambda_s + \mu_{sc})} e^{-(\lambda_s + \mu_{sc})t} \tag{13.25}$$

and

$$P_1(t) = \frac{\lambda_s}{(\lambda_s + \mu_{sc})} - \frac{\lambda_s}{(\lambda_s + \mu_{sc})} e^{-(\lambda_s + \mu_{sc})t} \tag{13.26}$$

The system availability and unavailability are given by

$$AV_s(t) = P_0(t) = \frac{\mu_{sc}}{(\lambda_s + \mu_{sc})} + \frac{\lambda_s}{(\lambda_s + \mu_{sc})} e^{-(\lambda_s + \mu_{sc})t} \tag{13.27}$$

$$UAV_s(t) = P_1(t) = \frac{\lambda_s}{(\lambda_s + \mu_{sc})} - \frac{\lambda_s}{(\lambda_s + \mu_{sc})} e^{-(\lambda_s + \mu_{sc})t} \tag{13.28}$$

where

$AV_s(t)$ is the system availability at time t and $UAV_s(t)$ is the system unavailability at time t.

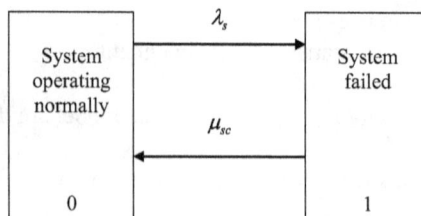

FIGURE 13.2   System-state space diagram.

As t becomes very large, Equation (13.27) and Equation (13.28) reduce to

$$AV_s = \frac{\mu_{sc}}{\left(\lambda_s + \mu_{sc}\right)} \tag{13.29}$$

and

$$UAV_s = \frac{\lambda_s}{\left(\lambda_s + \mu_{sc}\right)} \tag{13.30}$$

where

AV_s is the system steady-state availability, and UAV_s is the system steady-state unavailability.

Since $\lambda_s = \dfrac{1}{MTTF_s}$ and $\mu_{sc} = \dfrac{1}{CMT_{sm}}$, Equation (13.29) and Equation (13.30) become

$$AV_s = \frac{MTTF_s}{CMT_{sm} + MTTF_s} \tag{13.31}$$

and

$$UAV_s = \frac{CMT_{sm}}{CMT_{sm} + MTTF_s} \tag{13.32}$$

where
$MTTF_s$ is the system mean time to failure.
CMT_{sm} is the system mean corrective maintenance time.

Example 13.3

Assume that a system's mean time to failure is 1000 hours and its mean corrective maintenance time, or mean time to repair, is 20 hours. Calculate the system
steady-state unavailability if the system failure and corrective maintenance times
follow exponential distribution.
By substituting the given data values into Equation (13.32), we obtain

$$UAV_s = \frac{20}{20 + 1{,}000} = 0.0196$$

This means the system steady-state unavailability is 0.0196, or there is 1.96%
chance that the system will be unavailable for service.

13.10.2 Mathematical Model II

This mathematical model represents a parallel system made up of two identical units. For the successful operation of the system, at least one unit must operate normally. The system fails when both the units fail. Repair or corrective maintenance starts as soon a unit fails to return to its operating state. The system-state space diagram is shown in Figure 13.3. The numerals in boxes and rectangle denote system states.

The model is subject to the following four assumptions:

- The system contains two identical and independent units.
- Unit failure and repair or corrective maintenance rates are constant.
- The repaired unit is as good as new.
- No corrective maintenance or repair is conducted when both the units fail or the system fails.

The symbols used for developing equations for the model are defined below:

- k is the kth system state; $k = 0$ (both units are operating normally), $k = 1$ (one unit failed, other operating normally), $k = 2$ (both units failed).
- λ_u is the unit failure rate.
- μ_u is the unit repair or corrective maintenance rate.
- $P_k(t)$ is the probability that the system is in state k at time t, for $k = 0, 1, 2$.

Using the Markov method and Figure 13.3, we obtain the following equations [2, 18, 19]:

$$\frac{dP_0(t)}{dt} + 2\lambda_u P_0(t) = \mu_u P_1(t) \tag{13.33}$$

$$\frac{dP_1(t)}{dt} + (\mu_u + \lambda_u) P_1(t) = 2\lambda_u P_0(t) \tag{13.34}$$

$$\frac{dP_2(t)}{dt} = \lambda_u P_1(t) \tag{13.35}$$

At time $t = 0$, $P_0(0) = 1$ and $P_1(0) = P_2(0) = 0$.
Solving Equation (13.33) to Equation (13.35), we obtain

$$P_0(t) = \left[\frac{\lambda_u + \mu_u + N_1}{N_1 - N_2}\right] e^{N_1 t} - \left[\frac{\lambda_u + \mu_u + N_2}{N_1 - N_2}\right] e^{N_2 t} \tag{13.36}$$

$$P_1(t) = \left[\frac{2\lambda_u}{N_1 - N_2}\right] e^{N_1 t} - \left[\frac{2\lambda_u}{N_1 - N_2}\right] e^{N_2 t} \tag{13.37}$$

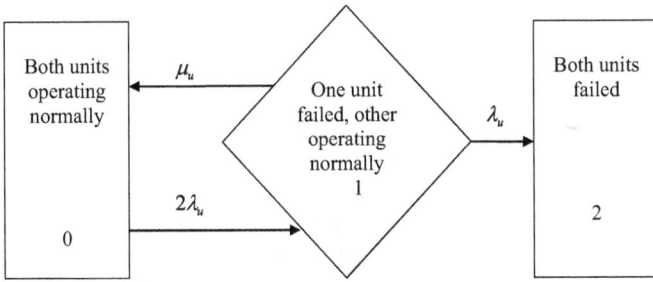

FIGURE 13.3 The two-unit parallel system state-space diagram.

and

$$P_2(t) = 1 + \left[\frac{N_2}{N_1 - N_2}\right]e^{N_1 t} - \left[\frac{N_1}{N_1 - N_2}\right]e^{N_2 t} \tag{13.38}$$

where

$$N_1, N_2 = \frac{-(3\lambda_u + \mu_u) \pm \left[(3\lambda_u + \mu_u)^2 - 8\lambda_u^2\right]^{1/2}}{2} \tag{13.39}$$

$$N_1 N_2 = 2\lambda_u^2 \tag{13.40}$$

and

$$N_1 + N_2 = -(3\lambda_u + \mu_u) \tag{13.41}$$

The parallel system reliability with repair is given by

$$R_{ps}(t) = P_0(t) + P_1(t) \tag{13.42}$$

where

$R_{ps}(t)$ is the parallel system reliability with repair at time t.

The parallel system mean time to failure with repair is given by

$$MTTF_{ps} = \int_0^\infty R_{ps}(t)\,dt$$
$$= \frac{3\lambda_u + \mu_u}{2\lambda_u^2} \tag{13.43}$$

where

$MTTF_{ps}$ is the parallel system mean time to failure with repair.

Example 13.4

Assume that an engineering system is composed of two identical and independent units, and at least one of the units must operate normally for system success. Both the units form a parallel configuration. A failed unit is repaired, but the failed system is never repaired. The unit times to failure and repair (i.e., corrective maintenance) are exponentially distributed and its failure and repair rates are 0.0008 failures per hour and 0.04 repairs per hour, respectively.

Calculate the system mean time to failure with and without the performance of corrective maintenance (i.e., repair) and comment on the final results.

By substituting the given data values into Equation (13.43), we obtain

$$MTTF_{ps} = \frac{3(0.0008)+0.04}{2(0.0008)^2}$$
$$= 33,125 \text{ hours}$$

By setting $\mu_n = 0$ in Equation (13.43) and substituting the given unit failure rate, we get

$$MTTF_{ps} = \frac{3}{2\lambda_u}$$
$$= \frac{3}{2(0.0008)}$$
$$= 1875 \text{ hours}$$

Thus, the system mean time to failure with and without the performance of corrective maintenance (i.e., repair) are 33,125 and 1875 hours, respectively. It means the performance of corrective maintenance (i.e., repair) on a unit has helped increase system mean time to failure from 1875 to 33,125 hours.

PROBLEMS

1. Define the terms: "Preventive maintenance" and "Corrective maintenance."
2. Discuss at least six elements of preventive maintenance.
3. Discuss steps for developing an effective preventive maintenance program in a short time.
4. Define the following two preventive maintenance measures:
   • Median preventive maintenance time.
   • Mean preventive maintenance time.

5. What are the advantages and disadvantages of performing preventive maintenance?
6. Discuss at least four types of corrective maintenance.
7. What are the major components of the corrective maintenance downtime?
8. Discuss strategies for reducing the system-level corrective maintenance time.
9. Define the following corrective maintenance measure:
   • Median active corrective maintenance time.

10. Assume that a system's mean time to failure is 1500 hours and its mean corrective maintenance time is 30 hours. Calculate the system steady-state availability if the system failure and corrective maintenance times follow exponential distribution.

REFERENCES

1. AMCP 706-132, *Engineering Design Handbook: Maintenance Engineering Techniques*, Department of Defense, Washington, DC, 1975.
2. Dhillon, B.S., *Engineering Maintenance: A Modern Approach*, CRC Press, Boca Raton, Florida, 2002.
3. Niebel, B.W., *Engineering Maintenance Management*, Marcel Dekker, New York, 1994.
4. McKenna, T., Oliverson, R., *Glossary of Reliability and Maintenance Terms*, Gulf Publishing, Houston, TX, 1997.
5. Omdahl, R.P., *Reliability, Availability, and Maintainability (RAM) Dictionary*, ASQC Quality Press, Milwaukee, WI, 1988.
6. Levitt, J., Managing Preventive Maintenance, *Maintenance Technology*, February 20–30, 1997.
7. Levitt, J., *Maintenance Management*, Industrial Press, New York, 1997.
8. Patton, J.D., *Preventive Maintenance*, Instrument Society of America, Research Triangle Park, NC, 1983.
9. Westerkemp, T.A., *Maintenance Manager's Standard Manual*, Prentice Hall, Paramus, NJ, 1997.
10. Blanchard, B.S., Verma, D., Peterson, E.L., *Maintainability*, John Wiley and Sons, New York, 1995.
11. Wild, R., *Essentials of Production and Operations Management*, Holt, Rinehart, and Winston, London, 1985.
12. Dhillon, B.S., *Mechanical Reliability: Theory, Models, Applications*, American Institute of Aeronautics and Astronautics, Washington, DC, 1988.
13. Dhillon, B.S., *Power System Reliability, Safety, and Management*, Ann Arbor Science Publishers, Ann Arbor, MI, 1983.
14. Dhillon, B.S., *Reliability Engineering in Systems Design and Operations*, Van Nostrand Reinhold Company, New York, 1983.
15. MICOM 705-8, *Maintenance of Supplies and Equipment*, Department of Defense, Washington, DC, 1972.
16. NAVORD OD 39223, *Maintainability Engineering Handbook*, Department of Defense, Washington, DC, June 1969.
17. AMCP-766-133, *Engineering Design Handbook: Maintainability Engineering Theory and Practice*, Department of Defense, Washington, DC, 1976.
18. Dhillon, B.S., *Design Reliability: Fundamentals and Applications*, CRC Press, Boca Raton, Florida, 1999.
19. Shooman, M.L., *Probabilistic Reliability: An Engineering Approach*, McGraw-Hill, New York, 1968.

14 Software and Robotic Maintenance

14.1 INTRODUCTION

Today, the maintenance field has developed to a level where it has started to branch out into many specialized areas including software maintenance and robotic maintenance.

Software maintenance is the process of making changes to the software system/component subsequent to delivery for improving performance or other attributes, rectifying faults, or adapting to a change in the use environment [1, 2]. In the early years of computing, software maintenance was only a small element of the overall software life cycle, but in recent years, it has become a major factor. For example, in 1955, the proportion of time spent on maintenance-related activities was around 23%; in 1970, it increased to around 36%, and the prediction for 1985 was around 58% [3–5]. As per Ref. [6], in the mid-1980s, the United States spent around $30 billion annually on software maintenance.

Although robots are generally quite reliable, sometimes, they do malfunction and require maintenance, just as is the case for any other sophisticated machines. Thus, the robots' users must devise effective maintenance programs; otherwise, their unscheduled downtime may increase to a point at which it defeats the purpose of robot applications. Moreover, careful consideration to maintenance should be given not only during the robots' operational phase but also during their design phase because various decisions concerning maintenance are made during this phase [7].

This chapter presents various important aspects of software and robotic maintenance.

14.2 SOFTWARE MAINTENANCE-RELATED FACTS AND FIGURES

There are many software maintenance-related facts and figures. Some of these are as follows:

- As per Ref. [8], over 80% of the life of a software product is spent in maintenance.
- The maintenance of existing software can consume over 60% of all development-associated efforts [9].
- Software maintenance activities account for around 70% of the overall software cost [9].
- A study conducted by the Boeing Company reported that annually, on average, 15% of the lines of source code in simple programs are changed, 5% are changed in medium programs, and 1% are changed in difficult programs [10].

DOI: 10.1201/9781003365204-14

- A study conducted by Hewlett-Packard reported that around 60-80% of its software research and development staff members are involved in maintenance of existing software [11].
- Extensions and modifications requested by users account for over two-fifths of software maintenance-related activities [9].
- It is estimated that for all software systems combined, the maintenance component of the overall effort is increasing around 3% annually [10].

14.3   SOFTWARE MAINTENANCE TYPES

Software maintenance may be broken down under the following four classifications [5, 12]:

- **Classification I: Preventive maintenance.** This is concerned with modifying software for enhancing potential reliability and maintainability or providing an improved basis for future enhancements.
- **Classification II: Perfect maintenance.** This is concerned with adding capabilities, modifying existing functions, and making general enhancements.
- **Classification III: Adaptive maintenance.** This is concerned with modifying software to effectively interface with a changing environment (i.e., both hardware and software).
- **Classification IV: Corrective maintenance.** This incorporates diagnosis and rectification of errors.

It is to be noted that a survey of 487 software organizations reported the percentage distribution of the above four types of maintenance as 4%, 50%, 25%, and 21%, respectively.

14.4   GUIDELINES FOR REDUCING SOFTWARE MAINTENANCE AND SOFTWARE MAINTENANCE TOOLS

There are many guidelines for reducing software maintenance. Nine of these guidelines are as follows [13–19]:

- **Guideline I:** Identify all possible software enhancements and design the software so that it can easily incorporate such enhancements.
- **Guideline II:** Carefully consider all human factors in areas that are the sources of frequent changes or modifications such as screen layouts.
- **Guideline III:** Employ preventive maintenance approaches such as using limits for tables that are reasonably greater than can possibly be needed.
- **Guideline IV:** Store constants in tables rather than scattering them throughout the software program.
- **Guideline V:** Introduce structured maintenance that uses methods for documenting existing systems, and incorporates guidelines for reading programs, etc.
- **Guideline VI:** Divide the functions into two groups: inherently more stable and most likely to be changed.

- **Guideline VII:** Use standard methodologies.
- **Guideline VIII:** Establish effective communication among maintenance programmers.
- **Guideline IX:** As much as possible, use portable languages, tools, and operating systems.

Over the years, many methods have been developed that directly or indirectly concern software maintenance [20]. Two of these methods are presented in the following two subsections:

14.4.1 SOFTWARE CONFIGURATION MANAGEMENT

During software maintenance, keeping track of changes and their effects on other system components is a quite challenging task, and software configuration management is an effective tool for meeting this challenge. Software configuration management is a set of tracking and control-related activities that start at the beginning of a software development project and terminate at the software retirement.

Configuration management is practiced by establishing a configuration control board because many software maintenance-related changes are requested by users for rectifying failures or making enhancements. The board oversees the change process, and its members are users, developers, and customers. Each problem is handled in the following seven ways [5, 12]:

- **Way I:** Software users, developers, or customers find a problem and use a formal change control form for recording all its associated symptoms and information.
- **Way II:** The proposed change is formally reported to the configuration control board.
- **Way III:** The board members discuss the proposed change.
- **Way IV:** The board makes a decision on the proposed change, prioritizes it, and assigns individuals for making the changes.
- **Way V:** These individuals identify the problem source, highlight the changes needed, and then test and implement the changes.
- **Way VI:** The designated people work along with the software program librarian for tracking and controlling the change in the operational system and updating related documentation.
- **Way VII:** The designated individual files the report describing the changes made.

Additional information on this method is available in Ref. [21].

14.4.2 IMPACT ANALYSIS

Software maintenance depends on and starts with user needs. Past experiences over the years clearly indicate that a need translating into a seemly minor change is often more extensive and costly than anticipated. Under such circumstance, impact analysis is a quite useful tool for determining the risks related to the proposed change, including the estimation of effects on factors such as resources, effort, and schedule. Various ways for measuring the impact of a given change are presented in Ref. [21].

14.5 SOFTWARE MAINTENANCE COST ESTIMATION MODELS

Over the years, many mathematical models have been developed for estimating various types of software maintenance cost. This section presents two such models that can directly or indirectly be used for estimating software maintenance cost.

14.5.1 MODEL I

This is a quite useful model for demonstrating how maintenance cost can build up alarmingly fast. The model is subject to the following two assumptions [22, 23]:

- **Assumption I:** The programming work force is constant and is normalized to be unity.
- **Assumption II:** After the completion of the project, a (normalized) maintenance force, m, is assigned for performing maintenance activities. Consequently, a (normalized) work force, b, is left for developing software for new projects.

Thus, we have

$$WF_p = m(t) + b(t) \tag{14.1}$$

where

WF_p is the programming work force.
$m(t)$ is the normalized maintenance work force at time t.
$b(t)$ is the normalized force left for developing software for new projects at time t.

From Equation (14.1), we note that at $t = 0$, $m = 0$ and $WF_p = b = 1$.
We define the fraction of the development force, y, assigned for maintenance at the completion of a software project as follows:

$$y = \frac{m}{WF_p} \tag{14.2}$$

However, at time $t = 0$, y is not defined.
If we start our first project at $t = 0$, then at its release, that is, $t = t_1$, we have

$$m(t_1) = yB(t_1) = y.1 = y \tag{14.3}$$

and

$$B = 1 - m = 1 - y \qquad (14.4)$$

After the release of the second project,

$$m = \text{assignment to project no.1} + \text{assignment to project no.2}$$
$$= y + yB \qquad (14.5)$$

By inserting Equation (14.4) into Equation (14.5), we get

$$m = y + y(1-y) \qquad (14.6)$$

and

$$B = WF_p - m$$
$$= 1 - m \qquad (14.7)$$

By substituting Equation (14.6) into Equation (14.7), we obtain

$$B = 1 - \left[y + y(1-y) \right]$$
$$= (1-y)^2 \qquad (14.8)$$

Similarly, using Equation (14.3) to Equation (14.8) for the nth release, we write

$$m = 1 - (1-y)^n \qquad (14.9)$$

and

$$B = (1-y)^n \qquad (14.10)$$

Example 14.1

Assume that after the completion of a software project, 10% of the work force is assigned to maintenance activities. There are a total of six projects of 4 years' duration. Estimate the percentage of the total work force that will be assigned to the maintenance activity of all the six projects.

By inserting the given data values into Equation (14.9) and Equation (14.10), we obtain

$$m = 1 - (1 - 0.10)^6$$
$$= 0.4658$$

and

$$B = (1-0.10)^6$$
$$= 0.5314$$

Thus, around 47% of the entire work force will be assigned to the maintenance aspect of all six software projects.

14.5.2 MODEL II

This is a quite useful model to estimate software maintenance cost. The software maintenance cost is defined by [24, 25]

$$MC_s = \left[3\left(C_{pm}\right)d\right]/\theta \qquad (14.11)$$

where

MC_s is the software maintenance cost.
C_{pm} is the cost per person-month.
d is the total number of instructions to be changed per month.
θ is the difficulty constant and its specified values are 500, 250, and 100 for easy, medium, and hard programs, respectively.

14.6   ROBOT MAINTENANCE REQUIREMENTS AND TYPES

Robot maintenance-related requirements are determined by the robot type and its application. Probably, the most important part that affects the need for maintenance and the provision of maintenance is the robot power system.

Most robots used in the industrial area can be grouped under the following two classifications [26]:

- **Classification I:** Electrical.
- **Classification II:** Hydraulic with electrical controls.

It is to be noted that irrespective of robot type, the mechanical components of robots require careful attention.

Maintenance of the robots being used in the industrial sector can be divided into the following three basic categories [27]:

- **Category I: Predictive maintenance.** This is concerned with predicting failures that may occur and alerting the appropriate maintenance personnel. Many robots are equipped with highly sophisticated electronic components and sensors.
- **Category II: Preventive maintenance.** This is concerned with the periodic servicing of robot system components.
- **Category III: Corrective maintenance.** This is concerned with repairing the robot to an operational state after its breakdown.

14.7 ROBOT PARTS AND TOOLS FOR MAINTENANCE AND REPAIR

A repair is composed of various parts, accessories, and subsystems. Thus, maintenance personnel must be familiar with such robot elements in order to conduct their tasks effectively. Nonetheless, 19 of these elements are as follows [28]:

- Element I: Hydraulic power supply
- Element II: Printed circuit board
- Element III: Alpha numeric keyboard
- Element IV: Cartesian coordinate system
- Element V: Strain gauge sensor
- Element VI: Cathode ray tube (CRT)
- Element VII: Mass memory device
- Element VIII: Servo valve
- Element IX: DC servomotor
- Element X: Proximity sensor
- Element XI: Bubble memory
- Element XII: Core memory
- Element XIII: Limit switch
- Element XIV: Air cylinder
- Element XV: Pressure transducer
- Element XVI: Stepping motor
- Element XVII: Microcomputer
- Element XVIII: Microprocessor
- Element XIX: Encoder

In performing robot maintenance, various types of tools are used, ranging from wrenches to diagnostic codes displayed on the robot control panel. Although the maintenance tools required are peculiar to the specific robot system in question, some of the most commonly used tools are shown in Figure 14.1 [29].

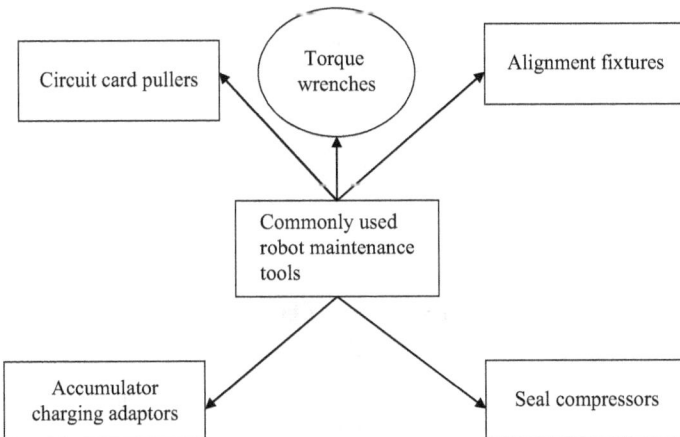

FIGURE 14.1 Some most commonly used robot maintenance tools.

14.8   ROBOT INSPECTION

Normally, robots are inspected regularly by their users. Nonetheless, the inspections of robots used in the industrial sector may be grouped into the following two broad categories [30]:

- **Category I:** Inspections performed prior to daily operations.
- **Category II:** Inspections performed at regular intervals.

In category I, some of the items checked prior to daily operations of the robots are as follows [30]:

- The proper working of interlocking the mechanism of associated items with the robot.
- The proper working of interlocking between the contact prevention equipment and the robot.
- The state of items used for the prevention of contact with the robot in operation.
- The proper functioning of the emergency stop.
- The presence of abnormality in the robot operation.
- The presence of abnormality in the robot supply air pressure.
- The proper working of the breaking device.
- The presence of abnormality in the robot supply oil pressure.
- The presence of abnormality in the supply voltage.
- The presence of abnormal noise.
- The presence of abnormal vibrations.
- Damage to external electric wires and piping.

Similarly, in category II, some of the items checked at regular intervals are as follows [30]:

- Abnormality in the operational troubleshooting function.
- Abnormal conditions in the electrical system.
- Abnormality in the air pressure system.
- The looseness of bolts in major robot parts.
- Abnormality in the lubrication of movable parts.
- Abnormality in the power train.
- Abnormality in the oil pressure system.
- Abnormality in the servo-system.
- Abnormal conditions in stoppers.
- Encoder abnormality.

14.9   USEFUL GUIDELINES FOR SAFEGUARDING ROBOT MAINTENANCE PERSONNEL

During robot maintenance, utmost care must be given for protecting robot maintenance and repair personnel. Four useful guidelines for this purpose are as follows [31]:

- **Guideline I:** Ensure that the robot system is properly switched off during all maintenance and repair activities as well as that the sources of power and the releasing of potentially dangerous stores energy are appropriately locked out or tagged.
- **Guideline II:** Ensure that all maintenance personnel have proper training in procedures appropriate to conduct the required tasks safely.
- **Guideline III:** Ensure that when a lockout or tag out procedure is not used, equally effective alternative safeguarding methods are employed.
- **Guideline IV:** Ensure that all maintenance personnel are properly protected from unexpected robot motion.

When it is not possible to turn off power during maintenance, six useful guidelines for protecting maintenance personnel are as follows [32]:

- **Guideline I:** Make the emergency stop readily accessible and make restarting the robot impossible until the emergency stop device is reset through manual means.
- **Guideline II:** Place robot controls in the hands of a second person who is knowledgeable regarding robot-related hazards as well as is capable of reacting fast to protect others in a moment of need.
- **Guideline III:** Place the robot arm in a predetermined position so that the required maintenance tasks can be conducted without exposing humans to trapping points.
- **Guideline IV:** Use devices such as pins and blocks during maintenance for preventing the robot system's potentially hazardous moments.
- **Guideline V:** Place the entire control of the robot in the hands of the maintenance person.
- **Guideline VI:** Reduce the robot speed to a slow speed level.

14.10   MODEL FOR MAXIMIZING INCOME OF ROBOT SUBJECT TO REPAIR

This model is concerned with maximizing the income of a robot subject to failure and repair. The robot availability and unavailability are expressed by [27, 33]

$$RAV(t) = \frac{\mu_r}{(\lambda_r + \mu_r)} + \frac{\lambda_r}{(\lambda_r + \mu_r)} e^{-(\lambda_r + \mu_r)t} \tag{14.12}$$

and

$$RUAV(t) = \frac{\lambda_r}{(\lambda_r + \mu_r)} \left[1 - e^{-(\lambda_r + \mu_r)t} \right] \tag{14.13}$$

where

> $RAV(t)$ is the robot availability at time t.
> $RUAV(t)$ is the robot unavailability at time t.
> λ_r is the robot constant failure rate.
> μ_r is the robot constant repair rate.

For a large value of time t, Equation (14.12) and Equation (14.13) simplify to

$$RAV = \frac{\mu_r}{(\lambda_r + \mu_r)} = \frac{RMTTF}{(RMTTF + RMTTR)} \tag{14.14}$$

and

$$RUAV = \frac{\lambda_r}{(\lambda_r + \mu_r)} = \frac{RMTTR}{(RMTTF + RMTTR)} \tag{14.15}$$

where

> RAV is the robot steady-state availability.
> $RUAV$ is the robot steady-state unavailability.
> $RMTTF$ is the robot mean time to failure.
> $RMTTR$ is the robot mean time to repair.

Equation (14.14) and Equation (14.15) may also be interpreted as the fraction of the time the robot repair crew is idle and the fraction of the time the robot crew is working, respectively.

The robot maintenance crew cost per month is expressed by

$$C_{rc} = \alpha \mu_r = \frac{\alpha}{RMTTR} \tag{14.16}$$

where

> C_{rc} is the robot maintenance crew cost per month.
> α is the robot maintenance cost constant that depends on the type of robot.

The expected income from the robot output per month is given by

$$
\begin{aligned}
IN_e &= IN_{rof} RAV \\
&= IN_{rof} \left[\frac{RMTTF}{(RMTTF + RMTTR)} \right]
\end{aligned} \tag{14.17}
$$

where

IN$_e$ is the expected income from the robot output per month.
IN$_{rof}$ is the income from the robot output per month if the robot worked full time.

Thus, the net income of the robot is

$$RNI = IN_e - C_{rc}$$
$$= IN_{rof} \left[\frac{RMTTF}{(RMTTF + RMTTR)} \right] - \frac{\alpha}{RMTTR} \tag{14.18}$$

To maximize the net income of the robot, we differentiate Equation (14.18) with respect to RMTTR and set the resulting derivatives equal to zero:

$$\frac{d(RNI)}{d(RMTTR)} = -IN_{rof} \left[\frac{RMTTF}{(RMTTF + RMTTR)^2} + \frac{\alpha}{(RMTTR)^2} \right] = 0 \tag{14.19}$$

By rearranging Equation (14.19), we obtain

$$RMTTR^* = \frac{RMTTF}{\left[\dfrac{IN_{rof}(RMTTF)}{\alpha} \right]^{1/2} - 1} \tag{14.20}$$

where

RMTTR* is the optimum value of RMTTR.

By inserting Equation (14.20) into Equation (14.14), Equation (14.16), and Equation (14.18), respectively, we obtain

$$RAV^* = 1 - \left[\frac{\alpha}{IN_{rof}(RMTTF)} \right]^{1/2} \tag{14.21}$$

$$C_{rc}^* = \left[\frac{\alpha IN_{rof}}{RMTTF} \right]^{1/2} - \frac{\alpha}{RMTTF} \tag{14.22}$$

and

$$RNI^* = IN_{rof} - \left[\frac{IN_{rof}\alpha}{RMTTF} \right] \tag{14.23}$$

where

RAV*, C_{rc}^*, and RNI* are the optimum values of RAV, C_{rc}, and RNI, respectively.

PROBLEMS

1. List at least five facts and figures directly or indirectly concerned with software maintenance.
2. What are the four classifications of software maintenance?
3. Discuss at least five guidelines for reducing software maintenance.
4. Describe the following method with respect to software maintenance:
    *   Software configuration management.
5. Assume that after the completion of a software project, 15% of the work force is assigned to maintenance activities. There are a total of seven projects of 4 years' duration. Estimate the percentage of the total work force that will be assigned to the maintenance activity of all the seven projects.
6. Define the formula/model that can be used to estimate software maintenance cost.
7. What are the basic categories of maintenance for industrial robots?
8. List at least 16 important robot parts.
9. List at least four most commonly used tools for robot maintenance.
10. Discuss at least five useful guidelines to protect maintenance personnel in situation when it is not possible to turn off power during maintenance.

REFERENCES

1. Omdahl, T.P. Ed., *Availability, and Maintainability (RAM) Dictionary*, ASQC Quality Press, Milwaukee, WI, 1988.
2. IEEE Standard Glossary of Software Engineering Technology, *IEEE-STD-610.12-1990*, Institute of Electrical and Electronic Engineers, New York, 1991.
3. Boem, B.W., *Software Engineering Economics*, Prentice Hall, Englewood Cliffs, NJ, 1981.
4. Stevenson, C., *Software Engineering Productivity*, Chapman and Hall, London, 1995.
5. Dhillon, B.S., *Maintainability, Maintenance, and Reliability for Engineers*, CRC Press, Boca Raton, Florida, 2006.
6. Martin, J., *Fourth-Generation Languages*, Vol. 1, Prentice Hall, Englewood Cliffs, NJ, 1985.
7. AMC Pamphlet No. *AMCP 706-133, Maintainability Engineering Theory and Practice, Prepared by Department of the Army*, Headquarter U.S. Material Command, Alexandria, VA, 1976.
8. Charrettee, R.N., *Software Engineering Environments*, Intertext Publications, New York, 1986.
9. Stacey, D., *Software Engineering, Course 27-320 Lecture Notes*, Department of Computer Science, University of Guelph, Guelph, Ontario, Canada, 1999.
10. Boeing Company, Software Cost Measuring and Reporting, ASD, Document No. D180-22813-1, U.S. Air Force, Washington, DC, 1979.
11. Coleman, D., Using Metrics to Evaluate Software System Maintainability, *Computer*, Vol. 27, No. 8, 1997, pp. 44–49.
12. Pfleeger, S.L., *Software Engineering Theory and Practice*, Prentice Hall, Upper Saddle River, NJ, 1998.
13. Arthur, L.J., *Software Evaluation: The Software Maintenance Challenge*, John Wiley and Sons, New York, 1983.

14. Parikh, G., Three Ts Keys to Maintenance Programming, *Computing S.A.*, April Vol. 22, pp. 19, 1981.
15. Hall, R.P., Seven Ways to Cut Software Maintenance Costs, *Datamation*, July 1987, pp. 81–87.
16. Schneider, G.R.E., Structural Software Maintenance, *Proceedings of the AFPIS National Computer Conference*, 1983, pp. 137–144.
17. Gilb, T., *Principles of Software Engineering Management*, Addison-Wesley, Wokingham, Berkshire, U.K., 1988.
18. Lindhorst, W.M., Scheduled Maintenance of Application Software, *Datamation*, May 1973, pp. 64–67.
19. Yourdon, E., Structured Maintenance, in *Techniques of Program and System Maintenance*, Parikh, G., Ed., Ethnotech, Lincoln, NE, 1980, pp. 211–213.
20. Holbrook, H.B., Thebaut, S.M., A Survey of Software Maintenance Tools that Enhance Program Understanding, Report No. SERC-TR-9-F, Software Engineering Research Center, Department of Science, Purdue University, West Lafayette, IN, 1987.
21. Dhillon, B.S., *Engineering Maintenance: A Modern Approach*, CRC Press, Boca Raton, FL, 2002.
22. Mills, M.D., Software Development, *Proceedings of the IEEE Second International Conference on Software Engineering*, Vol. 2, 1976, pp. 79–83.
23. Shooman, M.L., *Software Engineering*, McGraw-Hill, New York, 1983.
24. Sheldon, M.R., *Life Cycle Costing: A better Method of Government Procurement*, Westview Press, Boulder, CO, 1979.
25. Dhillon, B.S., *Life Cycle Costing*, Gordon and Breach Science Publishers, New York, 1989.
26. Lester, W.A., Lannon, R.P., Bellandi, R., Robot Users Needs to have a Program for Maintenance, *Industrial Engineering*, January 1985, pp. 28–32.
27. Dhillon, B.S., *Robot Reliability and Safety*, Springer Verlag, New York, 1991.
28. Ottinger, L.V., Robot System's Success Based on Maintenance, in *Robotics*, Fishers, E.L., Ed., Industrial Engineering and Management Press, Institute of Industrial Engineers, Atlanta, GA, 1983, pp. 204–208.
29. Munson, G.E., Industrial Robots: Reliability, Maintenance, and Safety, in *Handbook of Industrial Robotics*, Nof, S.Y., Ed., John Wiley and Sons, New York, 1985, pp. 722–758.
30. An Interpretation of the Technical Guidance on Safety Standards in the Use, etc., of Industrial Robots, *Japanese Ministry of Labor*, Japanese Industrial Safety and Health Association, Tokyo, 1985.
31. ANSI/RIA R15-6, *American National Standard for Industrial Robots and Robot Systems: Safety Requirements*, prepared by the Robotic Industries Association, Ann Arbor, MI, 1986.
32. Lodge, J.E., How to Protect Maintenance Workers, *National Safety News*, June 1984, pp. 48–51.
33. Dhillon, B.S., *Design Reliability: Fundamentals and Applications*, CRC Press, Boca Raton, Florida, 1999.

15 Reliability-Centered Maintenance

15.1 INTRODUCTION

Reliability-centered maintenance (RCM) systematically highlights the preventive maintenance-related tasks required for sustaining, in the most cost-effective manner possible, the maximum level of reliability and safety that can be expected from a product when it receives effective maintenance.

The history of RCM began within the commercial aircraft industrial sector in the late 1960s. A 1968 handbook titled "Maintenance Evaluation and Program Development," prepared by the United States Air Transport Association (ATA) for use with the Boeing 747 aircraft, contained one of the earliest formal treatments of the subject [1–3]. In 1970, a revised version of the handbook also discussed two other wide-body aircraft, the DC-10 and L-1011 [4, 5]. However, it was in 1974, when the United States Department of Defense commissioned United Airlines to prepare a document on civil aviation aircraft maintenance programs, that the term "reliability-centered maintenance" was coined as the title of the resulting document [6]. Additional information on the history of RCM is available in Ref. [5].

This chapter presents various important aspects of RCM.

15.2 RCM GOALS AND PRINCIPLES

There are many goals of RCM. Four important ones are as follows [7]:

- **Goal I:** To collect information considered useful for improving the design of items with proven unsatisfactory inherent reliability.
- **Goal II:** To develop preventive maintenance-related tasks that can effectively reinstate safety and reliability to their inherent levels in the event of system/equipment deterioration.
- **Goal III:** To develop design-associated priorities that can effectively facilitate preventive maintenance.
- **Goal IV:** To achieve all the previously mentioned goals when the total cost is minimal.

There are many principles of RCM. Ten of these principles are as follows [8]:

- **Principle I: Three types of maintenance tasks along with run-to-failure tasks are acknowledged by RCM.** These tasks are known as failure finding, condition directed, and time directed. The purpose of failure-finding tasks is to find hidden functions that have failed without providing any

DOI: 10.1201/9781003365204-15

indication of pending failure. Condition-directed tasks are conducted as the indications indicate for their necessity. Time-directed tasks are scheduled as considered appropriate. Run-to-failure is a conscious decision in RCM.

- **Principle II: Design-associated limitations are acknowledged by RCM.** It means that the goal of RCM is to maintain the inherent reliability of the equipment/system design and, at the same time, recognize that any changes in inherent reliability can only be made through design rather than maintenance. More specifically, maintenance at the best of times can only achieve and maintain at a level of designed reliability.
- **Principle III: RCM is reliability centered.** It means that RCM is not overly concerned with simple failure rate, but it places importance on the relationship between failures experienced and operating age. In short, RCM treats failure statistics in an actuarial fashion.
- **Principle IV: RCM is equipment/system focused.** It means that RCM is more concerned with maintaining system function as opposed to maintaining function of individual component.
- **Principle V: Safety and economics drive RCM.** It means that safety is of paramount importance; thus, it must be ensured properly at any cost, and then, cost-effectiveness becomes the criterion.
- **Principle VI: RCM is function oriented.** It means that RCM plays an instrumental role in preserving system/equipment function, not just operability for its own sake.
- **Principle VII: An unsatisfactory condition is defined as a failure by RCM.** It means that a failure could be either a loss of acceptable quality or a loss of function.
- **Principle VIII: RCM uses a logic tree for screening maintenance tasks.** This provides consistency in the maintenance of all types of equipment.
- **Principle IX: RCM tasks must be effective.** It means that the tasks must be technically sound and cost-effective.
- **Principle X: RCM tasks must be applicable.** It means that tasks must reduce failures or ameliorate secondary damage resulting from failure.

15.3 RCM PROCESS

The process is applied for determining specific maintenance tasks to be conducted as well as for influencing item reliability and maintainability during design. Initially, the RCM process is applied during the design and development phase and then reapplied, as necessary, during the operational phase to sustain an effective maintenance program based on experience in the field environment.

The basic RCM process is composed of the following seven steps [9]:

- **Step I: Highlight important items in regard to maintenance.** Usually, maintenance-important items are highlighted by using methods such as failure, mode, effects, and criticality analysis (FMECA) and fault tree analysis (FTA).

- **Step II: Obtain essential failure data.** In determining occurrence probabilities and assessing criticality, the availability of data on items such as operator error probability, part failure rate, and inspection efficiency is very important. These types of data come from generic failure data banks, field experience, etc.
- **Step III: Develop fault tree analysis (FTA) data.** The occurrence probabilities of fault events (i.e., basic, intermediate, and top events) are estimated as per combinatorial properties of the logic elements in the fault tree.
- **Step IV: Apply decision logic to critical failure modes.** The decision logic is designed to lead, by asking standard assessment-associated questions, to the most desirable task combinations of preventive maintenance. The same logic is applied to each and every crucial failure mode of each maintenance-important item.
- **Step V: Categorize maintenance requirements.** Maintenance-associated requirements are classified under three categories. These categories are hard-time maintenance requirements, condition monitoring maintenance requirements, and on-condition monitoring maintenance requirements.
- **Step VI: Implement RCM decisions.** Tasks intervals and frequencies are enacted/set as part of the overall maintenance plan/strategy.
- **Step VII: Apply sustaining engineering on the basis of real-life field experience.** Once system/equipment starts operating, the real-life field data start to accumulate. At that time, one of the most important steps is to re-evaluate all RCM associated default decisions.

15.4   RCM ELEMENTS

There are following four major elements of RCM [10–12]:

- Proactive maintenance
- Preventive maintenance
- Reactive maintenance
- Predictive testing and inspection

All of the above four major elements are described in the following four subsections:

15.4.1   PROACTIVE MAINTENANCE

Proactive maintenance is quite useful for improving maintenance through actions such as better design, workmanship, installation, maintenance procedures, and scheduling. The characteristics of proactive maintenance include items listed below [10]:

- Using feedback and communications for ensuring that changes in design/procedures are efficiently made available to item management/designers.
- Ensuring that nothing that affects maintenance occurs in total isolation, with the ultimate goal of correcting the concerned equipment forever.

- Optimizing and tailoring maintenance methods and technologies to each application.
- Practicing a continuous process of improvement.

Proactive maintenance integrates functions with support maintenance into maintenance program planning, conducts root cause failure analysis and predictive analysis for enhancing maintenance effectiveness, uses a life cycle view of maintenance and supporting functions, and conducts periodic evaluation of the technical content and the performance interval of maintenance tasks [12]. The following eight methods are used by proactive maintenance for extending equipment/item life [10, 12]:

- Failed item analysis
- Root cause failure analysis
- Reliability engineering
- Age exploration (AE)
- Recurrence control
- Precision rebuild and installation
- Specifications for new/rebuilt item/equipment
- Rebuild certification/verification

The first four of the above eight methods are described in the following four subsections and the information on the remaining four methods is available in Ref. [10].

15.4.1.1 Failed Item Analysis

Failed item analysis involves visually inspecting failed items after removal to determine failure causes. As the need arises, more detailed technical analysis is carried out for finding the real cause of a failure. For example, in the case of bearings, the root causes for the occurrence of their failures may relate to factors such as improper lubrication practices, poor storage and handling methods, excessive balance and alignment tolerances, or poor installation.

Past experiences, over the years, clearly indicate that over 50% of all bear-associated problems are due to improper installation or contamination. Usually, indicators of improper installation-associated problems are evident on bearings' internal and external surfaces and the indicators of contamination appear on the bearings' internal surfaces.

15.4.1.2 Root Cause Failure Analysis

Root cause failure analysis is concerned with proactively seeking the fundamental causes of equipment/facility failure. Four main objectives of conducting root cause failure analysis are as follows:

- **Objective I:** To provide data that can be useful for eradicating the problem.
- **Objective II:** To determine the cause of a problem economically and efficiently.
- **Objective III:** To instill a mentality of "fix forever."
- **Objective IV:** To rectify the problem cause, not just its effect.

15.4.1.3 Reliability Engineering

Reliability Engineering, in conjunction with other proactive maintenance methods, involves the modification, the improvement, or the redesign of items/parts or their replacement with better parts/items. In certain cases, a total redesign of the part/item under consideration may be required. There are many methods used in reliability engineering to perform reliability analysis of engineering systems/items. The two most widely used methods in the industry are failure modes and effect analysis (FMEA) and FTA.

Additional information on FMEA and FTA is available in Chapter 4 and in Ref. [5].

15.4.1.4 Age Exploration (AE)

AE is a very important factor in developing a RCM program. It provides a mechanism for varying key aspects of a maintenance program for optimizing the process. The AE method examines the applicability of all maintenance-related tasks in regard to the following three factors:

- **Factor I: Technical content.** The technical contents of a task are examined for ensuring that all highlighted modes of failures are appropriately addressed, as well as assuring that the current maintenance-related tasks result in the expected degree of reliability.
- **Factor II: Task grouping.** Tasks with similar periodicity are grouped or categorized for the purpose of improving the total time spent on the job site as well as reducing outages.
- **Factor III: Performance interval.** Adjustments are conducted continually to the interval of task performance until the rate at which resistance to failure declines is effectively sensed or determined.

15.4.2 PREVENTIVE MAINTENANCE (PM)

PM is also known as interval-based or time-driven maintenance and is conducted with respect to equipment condition. It consists of periodically scheduled inspection, calibration, lubrication, cleaning, adjustments, part replacement, and repair of components/items. PM schedules regular maintenance and inspection at set intervals for reducing failures for susceptible system/equipment.

It is to be noted that, depending on the predefined intervals, practicing PM can lead to a significant increase in routine inspections and maintenance. However, on the other hand, it can help to reduce the frequency and the severity of unplanned failures. If PM is the only type of maintenance practiced, it can be quite costly and ineffective. Additional information on PM is available in Ref. [10].

15.4.3 REACTIVE MAINTENANCE

This maintenance is also known as repair, run-to-failure, breakdown, or fix-when-fail maintenance. In this type of maintenance, it is assumed that there is an equally likely chance for a failure occurrence in any system, part, or component. When using this maintenance approach, system/equipment maintenance, repair, or replacement takes place only when deterioration in the condition of an equipment/item/system result in a functional failure.

When reactive maintenance is practiced solely, a high replacement of part inventories, a high percentage of unplanned activities, and poor use of maintenance-associated effort are usually typical. In addition, a totally reactive maintenance program overlooks opportunities for influencing system/equipment/item survivability.

It is to be noted that reactive maintenance can be practiced effectively only if it is conducted as a conscious decision, based on the conclusions of an RCM analysis that compares task and cost of failure with the cost of maintenance needed for mitigating that risk and failure cost. A criterion for determining the priority of repairing or replacing the failed item/equipment in the reactive maintenance program is available in Refs. [10, 12].

15.4.4   Predictive Testing and Inspection

Predictive testing and inspections (PTIs) are also referred to as predictive maintenance or condition monitoring. In order to assess item/equipment condition, it uses performance-related data, visual inspection, and nonintrusive testing methods. PTI replaces arbitrarily timed maintenance tasks with maintenance and is conducted as warranted by the item/equipment condition. Analysis of equipment/item condition monitoring data on a continuous basis is quite useful to plan and schedule maintenance/repair in advance of catastrophic/functional failure.

The collected (PTI) data are used for determining the equipment condition and for identifying the failure precursors in many ways, including trend analysis, pattern recognition, data comparison, statistical process analysis, correlation of multiple technologies, and tests against limits and ranges.

Finally, it is to be noted with care that PTI should not be the only type of maintenance practiced, because it does not lend itself to all types of item/equipment or possible failure modes.

15.5   RCM APPLICATION ADVANTAGES AND REASONS FOR ITS FAILURES

There are many advantages of the RCM methodology application. Six of these advantages are as follows [6]:

- **Advantage I:** Improvement in safety and environmental protection.
- **Advantage II:** Improvement in maintenance cost-effectiveness.
- **Advantage III:** Improvement in operating performance.
- **Advantage IV:** Improvement in individual motivation.
- **Advantage V:** Improvement in teamwork.
- **Advantage VI:** A maintenance database.

Although the application of RCM methodology can generate fast results and many advantages if it is applied effectively, but not every application of RCM may yield its full potential. In fact, some may achieve very little or nothing. Some of the reasons for this may be that the application was hurried or superfluous, the analysis was performed at too low a level, or that too much emphasis was placed on failure data, such as mean time to failure (MTTF) and mean time to repair (MTTR) [6].

15.6 RCM PROGRAM EFFECTIVENESS MEASUREMENT INDICATORS

In order to measure an RCM program's effectiveness, over the years, many management indicators have been developed. The numerical indicators/metrics are considered quite helpful because they are objective, quantitative, more easily trended than words and precise, in addition to consisting of a descriptor and a benchmark. A benchmark is a numerical expression of a set goal, and a descriptor may be expressed as a word or a group of words detailing the function, the units, as well as the process under consideration for measurement.

Six indicators/metrics considered quite useful for measuring the effectiveness of a RCM program are presented in the following six subsections along with their suggested benchmark values [10, 12]. These benchmark values are the mean values of data surveyed from around 50 major corporations in the early 1990s [12].

15.6.1 INDICATOR I: EQUIPMENT AVAILABILITY

Equipment availability is defined by

$$EAV = \frac{H_{eue}}{H_t} \tag{15.1}$$

where

EAV is the equipment availability.
H_{eue} is the number of hours that each unit of equipment is available to run at capacity.
H_t is the total number of hours during the reporting period.

For this indicator, the benchmark figure is 96%.

15.6.2 INDICATOR II: EMERGENCY PERCENTAGE INDEX

Emergency Percentage Index is defined by

$$EP = \frac{H_{tej}}{H_{tw}} \tag{15.2}$$

where

EP is the emergency percentage.
H_{tw} is the total number of hours worked.
H_{tej} is the total number of hours worked on emergency jobs.

For this indicator, the benchmark figure is 10% or less.

15.6.3 INDICATOR III: PREDICATIVE TESTING AND INSPECTION (PTI)-COVERED EQUIPMENT INDEX

This index is a metric that is used for calculating the percentage of candidate equipment covered by PTI and is expressed by

$$PCE = \frac{EIPP_t}{ECP_t} \qquad (15.3)$$

where

PCE is the percentage of candidate equipment covered by PTI.
$EIPP_t$ is the total number of equipment items in PTI program.
ECP_t is the total number of equipment candidates for PTI.

For this indicator, the benchmark figure is 100%.

15.6.4 INDICATOR IV: MAINTENANCE OVERTIME PERCENTAGE INDEX

Maintenance overtime percentage index is expressed by

$$MOP = \frac{H_{tmo}}{H_{trm}} \qquad (15.4)$$

where

MOP is the maintenance overtime percentage.
H_{tmo} is the total number of maintenance overtime hours during period.
H_{trm} is the total number of regular maintenance hours during period.

For this indicator, the benchmark figure is 5% or less.

15.6.5 INDICATOR V: PREVENTIVE MAINTENANCE (PM)/PTI-REACTIVE MAINTENANCE INDEX

PM/PTI-reactive maintenance index is an indicator that is divided into two areas: PM/PTI and reactive maintenance. The PM/PTI-associated index is expressed by

$$WP_{pp} = \frac{HH_{tpw}}{HH_{tpw} + HH_{trw}} \qquad (15.5)$$

where

WP_{pp} is the PM/PTI work percentage.
HH_{tpw} is the total human-hours of PM/PTI work.
HH_{trw} is the total human-hours of reactive maintenance work.

For this indicator, the benchmark figure is 70%.

The reactive maintenance-associated index is defined by

$$WP_{rm} = \frac{HH_{trw}}{HH_{trw} + HH_{tpw}}$$ (15.6)

where

WP_{rm} is the reactive maintenance work percentage.

For this indicator, the benchmark figure is 30%. It is to be noted that the sum of Equation (15.5) and Equation (15.6) is equal to unity or 100%.

15.6.6 INDICATOR VI: EMERGENCY-PM/PTI WORK INDEX

This index is expressed by

$$PEW_{pp} = \frac{H_{tew}}{H_{tppw}}$$ (15.7)

where

PEW_{pp} is the percentage of emergency work to PTI and PM work.
H_{tew} is the total number of emergency work hours.
H_{tppw} is the total number of PTI and PM work hours.

For this indicator, the benchmark figure is 20% or less.

PROBLEMS

1. What are the important goals of RCM?
2. What are the principles of RCM?
3. Describe the RCM process.
4. What are the major elements of RCM?
5. List at least eight methods that can be used by proactive maintenance for extending equipment/item life.
6. Describe the following two methods:
   - Failed item analysis
   - Root cause failure analysis
7. What are the advantages of RCM?
8. What are the reasons for the failures of RCM?
9. Define the following two indicators that can be used to measure effectiveness of a RCM program:
   - Emergency percentage index
   - Equipment availability
10. What are the benchmark figures for the problem 9 indicators?

REFERENCES

1. Anderson, R.T., Neri, L., *Reliability Centered Maintenance: Management and Engineering Methods*, Elsevier Applied Science Publishers, London, 1990.
2. MSGI, *Maintenance Evaluation and Program Development: 747 Maintenance Steering Group Handbook*, Air Transport Association, Washington, DC, 1968.
3. Smith, A.M., Vasudevan, R.V., Matteson, T.D., Gaertner, J.P., Enhancing Plant Preventive Maintenance Via RCM, *Proceedings of the Annual Reliability and Maintainability Symposium*, 1986, pp. 120–125.
4. MSG2, *Airline/Manufacturer Maintenance Program Planning Document*, Air Transport Association, Washington, DC, 1970.
5. Dhillon, B.S., *Engineering Maintainability*, Gulf Publishing Company, Houston, Texas, 1999.
6. Moubray, J., *Reliability-Centered Maintenance*, Industrial Press, Inc., New York, 1992.
7. AMC Pamphlet No. 750-2, *Guide to Reliability-Centered Maintenance*, Department of the Army, Washington, DC, 1985.
8. Nowlan, F.S., Heap, H.F., *Reliability Centered Maintenance*, Dolby Access Press, San Francisco, 1978.
9. Brauer, D.C., Brauer, G.D., Reliability-Centered Maintenance, *IEEE Transactions on Reliability*, Vol. 36, 1987, pp. 17–24.
10. Dhillon, B.S., *Engineering Maintenance: A Modern Approach*, CRC Press, Boca Raton, Florida, 2002.
11. Report No. NAVIR 00-25-403, *Guidelines for the Naval Aviation Reliability-Centered Maintenance Process, Naval Air Systems Command*, Department of Defense, Washington, DC, October 1996.
12. Reliability Centered Maintenance Guide for Facilities and Collateral Equipment, National Aeronautics and Space Administration, Washington, DC, 1996.

Index

For Product Safety Concerns and Information please contact our EU
representative GPSR@taylorandfrancis.com
Taylor & Francis Verlag GmbH, Kaufingerstraße 24, 80331 München, Germany